DMV Seminar
Band 25

Lectures on Spaces of Nonpositive Curvature

with an appendix by Misha Brin
Ergodicity of Geodesic Flows

Werner Ballmann

Birkhäuser Verlag
Basel · Boston · Berlin

Author:

Werner Ballmann
Mathematisches Institut der
Universität Bonn
Wegelerstr. 10
D-53115 Bonn

A CIP catalogue record for this book is available from the Library of Congress, Washington D.C., USA

Deutsche Bibliothek Cataloging-in-Publication Data
Ballmann, Werner:
Lectures on spaces of nonpositive curvature / Werner
Ballmann. With an appendix Ergodicity of geodesic flows / by
Misha Brin. - Basel ; Boston ; Berlin : Birkhäuser, 1995
(DMV-Seminar ; Bd. 25)
ISBN 3-7643-5242-6 (Basel ...)
ISBN 0-8176-5242-6 (Boston)
NE: Brin, Misha: Ergodicity of geodesic flows; Deutsche Mathematiker-
Vereinigung: DMV-Seminar

Camera-ready copy prepared by the author
Printed on acid-free paper produced from chlorine-free pulp ∞
Cover design: Heinz Hiltbrunner, Basel
Printed in Germany
ISBN 3-7643-5242-6
ISBN 0-8176-5242-6

9 8 7 6 5 4 3 2 1

Contents

Dedicated to my Mother

Introduction

These notes grew out of lectures which I gave at a DMV-seminar in Blaubeuren, Bavaria. My main aim is to present a proof of the rank rigidity for manifolds of nonpositive sectional curvature and finite volume. Since my interest in the last couple of years has shifted to singular spaces of nonpositive curvature, I take the opportunity to include a short introduction into the theory of these spaces. An appendix on the ergodicity of geodesic flows has been contributed by Misha Brin.

Let X be a metric space with metric d. A *geodesic* in X is a curve of constant speed which is locally minimizing. We say that X has *nonpositive Alexandrov curvature* if every point $p \in X$ has a neighborhood U with the following properties:

(i) for any two points $x, y \in U$ there is a geodesic $\sigma_{x,y} : [0,1] \to U$ from x to y of length $d(x, y)$;

(ii) for any three points $x, y, z \in U$ we have

$$d^2(z, m) \leq \frac{1}{2}\left(d^2(z,x) + d^2(z,y)\right) - \frac{1}{4}d^2(x,y)$$

where $\sigma_{x,y}$ is as in (i) and $m = \sigma_{x,y}(1/2)$ is in the middle between x and y. The second assumption requires that triangles in U are not fatter than the corresponding triangles in the Euclidean plane $\mathbb{R}^2$: if U is in the Euclidean plane then we have equality in (ii). It follows that the geodesic $\sigma_{x,y}$ in (i) is unique.

Modifying the above definition by comparing with triangles in the model surface M^2_κ of constant Gauss curvature κ instead of $\mathbb{R}^2 = M^2_0$, one obtains the definition of *Alexandrov curvature* $K_X \leq \kappa$. It is possible, but not very helpful, to define K_X and not only the latter inequality. Reversing the inequality in the triangle comparison, that is, requiring that (small) triangles are at least as fat as their counterparts in the model surface M^2_κ, leads to the concept of spaces with lower curvature bounds. The theory of spaces with lower curvature bounds is completely different from the theory of those with upper curvature bounds.

Triangle comparisons are a standard tool in global Riemannian geometry. In the Riemannian context our requirement (ii) is equivalent to nonpositive sectional curvature and an upper bound κ for the sectional curvature respectively.

We say that X is a *Hadamard space* if X is complete and the assertions (i) and (ii) above hold for all points $x, y, z \in X$. This corresponds to the terminology in the

Riemannian case. If X is a Hadamard space, then for any two points in X there is a unique geodesic between them. It follows that Hadamard spaces are contractible, one of the reasons for the interest in these spaces. There are many examples of Hadamard spaces, and one of the sources is the following result, Gromov's version of the Hadamard-Cartan theorem for singular spaces.

Theorem A (Hadamard-Cartan Theorem). *Let Y be a complete connected metric space of nonpositive Alexandrov curvature. Then the universal covering space X of Y, with the induced interior metric, is a Hadamard space.*

For X and Y as in Theorem A, let Γ be the group of covering transformations of the projection $X \to Y$. Since X is contractible, Y is the classifying space for Γ. This implies for example that the homology of Γ is equal to the homology of Y. The contractibility of X also implies that the higher homotopy groups of Y are trivial and that indeed the homotopy type of Y is determined by Γ. This is the reason why spaces of nonpositive curvature are interesting in topology. Since Γ acts isometrically on X, the algebraic structure of Γ and the homotopy type of Y are tied to the geometry of X.

The action of Γ on X is properly discontinuous and free, but for various reasons it is also interesting to study more general actions. This is the reason that we formulate our results below for groups Γ of isometries acting on Hadamard spaces instead of discussing spaces covered by Hadamard spaces.

Here are several examples of Hadamard spaces and metric spaces of nonpositive and respectively bounded Alexandrov curvature.

(1) Riemannian manifolds of nonpositive sectional curvature: the main examples are symmetric spaces of noncompact type, see [Hel, ChEb]. One particular such space is $X = Sl(n,\mathbb{R})/SO(n)$ endowed with the metric which is induced by the Killing form. Every discrete (or not discrete) linear group acts on this space (where n has to be chosen appropriately).

Many Riemannian manifolds of nonpositive curvature are obtained by using warped products, see [BiON].

For Riemannian manifolds with boundary there are conditions on the second fundamental form of the boundary – in addition to the bound on the sectional curvature – which are equivalent to the condition that the Alexandrov curvature is bounded from above, see [ABB].

(2) Graphs: Let X be a graph and d an interior metric on X. Then for an arbitrary $\kappa \in \mathbb{R}$, the Alexandrov curvature of d is bounded from above by κ if and only if for every vertex v of X there is a positive lower bound on the length of the edges adjacent to v.

This example shows very well one of the main technical problems in the theory of metric spaces with Alexandrov curvature bounded from above – namely the possibility that geodesics branch.

(3) Euclidean buildings of Bruhat and Tits: these are higher dimensional versions of homogeneous trees, and they are endowed with a natural metric (determined up to a scaling factor) with respect to which they are Hadamard spaces.

Similarly, Tits buildings of spherical type are spaces of Alexandrov curvature ≤ 1. See [BruT, Ti2, Bro].

(4) (p,q)-spaces with $p,q \geq 3$ and $2pq \geq p+q$: by definition, a two dimensional CW-space X is a (p,q)-space if the attaching maps of the cells of X are local homeomorphisms and if

(i) every face of X has at least p edges in its boundary (when counted with multiplicity);

(ii) for every vertex v of X, every simple loop in the link of v consisits of at least q edges.

If X is a (p,q)-space, then the interior metric d, which turns every face of X into a regular Euclidean polygon of side length 1, has nonpositive Alexandrov curvature.

In combinatorial group theory, (p,q)-spaces arise as Seifert-van Kampen diagrams or Cayley complexes of small cancellation groups, see [LySc, GeSh, BB2]. The angle measurement with respect to the Euclidean metric on the faces of X induces an interior metric on the links of the vertices of X, and the requirement $2pq \geq p+q$ implies that simple loops in the links have length at least 2π. This turns the claim that d has nonpositive Alexandrov curvature into a special case of the corresponding claim in the following example.

If $2pq > p+q$, then X admits a metric of Alexandrov curvature ≤ -1.

(5) Cones: Let X be a metric space. Define the *Euclidean cone* C over X to be the set $[0,\infty) \times X$, where we collapse $\{0\} \times X$ to a point, endowed with the metric

$$d_C((a,x),(b,y)) \;:=\; a^2 + b^2 - 2ab\cos(\min\{d_X(x,y),\pi\})\,.$$

Then C has nonpositive Alexandrov curvature if and only if X has Alexandrov curvature $K_X \leq 1$ and injectivity radius $\geq \pi$. By the latter we mean that for every two points $x,y \in X$ of distance $< \pi$ there is one and only one geodesic from x to y of length $d(x,y)$. In a similar way one can define the spherical and respectively hyperbolic cone over X. The condition on X that they have Alexandrov curvature ≤ 1 and respectively ≤ -1 remains the same as in the case of the Euclidean cone, see [Ber2, Gr5, Ba5, BrHa].

(6) Glueing: Let X_1 and X_2 be complete spaces of nonpositive Alexandrov curvature, let $Y_1 \subset X_1$ and $Y_2 \subset X_2$ be closed and locally convex and let $f : Y_1 \to Y_2$ be an isometry. Le X be the disjoint union of X_1 and X_2, except that we identify Y_1 and Y_2 with respect to f. Then X, with the induced interior metric, is a complete space of nonpositive curvature.

For X_1, Y_1 given as above one may take $X_2 = Y_1 \times [0,1]$, $Y_2 = Y_1 \times \{0\}$ and f the identity, that is, the map that forgets the second coordinate 0. One obtains a porcupine. Under the usual circumstances it is best to avoid such and similar exotic animals. One way of doing this is to require that X is *geodesically complete*, meaning that every geodesic in X is a subarc of a geodesic which is parameterized on the whole real line. Although the geodesic completeness does not exclude porcupines with spines of infinite length, it is a rather convenient regularity assumption and should be sufficient for most purposes, at least in the locally compact case. Of course one might wonder how serious such a restriction is and how many interesting examples are excluded by it. A rich source of examples are locally finite polyhedra with piecewise smooth metrics of nonpositive Alexandrov curvature and, in dimension 2, such a polyhedron always contains a homotopy equivalent subpolyhedron which is geodesically complete and of nonpositive Alexandrov curvature with respect to the induced length metric, see [BB3]. Whether this or something similar holds in higher dimensions is open (as far as I know).

In Riemannian geometry, the difference between strict and weak curvature bounds is very important and has attracted much attention. It is the contents of rigidity theorems that certain properties, known to be true under the assumption of a (certain) strict curvature bound, do not hold under the weak curvature bound, but fail to be true only in a very specific way. Well known examples are the Maximal Diameter Theorem of Toponogov and the Minimal Diameter Theorem of Berger (see [ChEb]). In the realm of nonpositive curvature, the above mentioned rank rigidity is one of the examples.

Let X be a Hadamard space. A *k-flat* in X is a convex subset of X which is isometric to Euclidean space $\mathbb{R}^k$. Rigidity phenomena in spaces of nonpositive curvature are often caused by the existence of flats. To state results connected to rank and rank rigidity, assume now that M is a smooth complete Riemannian manifold of nonpositive sectional curvature. Let $\gamma : \mathbb{R} \to M$ be a unit speed geodesic. Let $\mathcal{J}^p(\gamma)$ be the space of parallel Jacobi fields along γ and set

$$\operatorname{rank}(\gamma) \;=\; \dim \mathcal{J}^p(\gamma) \quad \text{and} \quad \operatorname{rank} M \;=\; \min \operatorname{rank}(\gamma)\,,$$

where the minimum is taken over all unit speed geodesics in M. Then

$$1 \;\le\; \operatorname{rank} M \;\le\; \operatorname{rank}(\gamma) \;\le\; \dim M$$

and

$$\operatorname{rank}(M_1 \times M_2) \;=\; \operatorname{rank} M_1 + \operatorname{rank} M_2\,.$$

The space $\mathcal{J}^p(\gamma)$ can be thought of as the maximal *infinitesimal flat* containing the unit speed geodesic γ. In general, $\mathcal{J}^p(\gamma)$ is not tangent to a flat in M. However, if M is a Hadamard manifold of rank k, then every geodesic of M is contained in a k-flat [B.Kleiner, unpublished]. (We prove this in the case that the isometry group of M satisfies the duality condition, see below.) If M is a symmetric space of noncompact type, then $\operatorname{rank} M$ coincides with the usual rank of M and is given as

$$\begin{aligned} \operatorname{rank} M \;&=\; \min\{k \mid \text{every geodesic of } M \text{ is contained in a } k\text{-flat}\} \\ &=\; \max\{k \mid M \text{ contains a } k\text{-flat}\}\,. \end{aligned}$$

The second property is false for Hadamard manifolds. Counterexamples are easy to construct.

Let X be a geodesically complete Hadamard space. The *geodesic flow* g^t, $t \in \mathbb{R}$, of X acts by reparameterization on the space GX of complete unit speed geodesics on X,

$$g^t(\gamma)(s) \;=\; \gamma(s+t)\,, \quad s \in \mathbb{R}\,.$$

We say that a unit speed geodesic $\gamma : \mathbb{R} \to X$ is *nonwandering* mod Γ if there are sequences of unit speed geodesics $(\gamma_n : \mathbb{R} \to X)$, real numbers (t_n) and isometries (φ_n) of X such that $\gamma_n \to \gamma$, $t_n \to \infty$ and $\varphi_n \circ g^{t_n}(\gamma_n) \to \gamma$. This corresponds to the usual definition of nonwandering in X/Γ in the case where Γ is a group of covering transformations and where geodesics in X/Γ are considered. In a similar way we translate other definitions which refer, in their standard versions, to objects in a covered space.

Following Chen and Eberlein [CE1] we say that a group Γ of isometries of X satisfies the *duality condition* if any unit speed geodesic of X is *nonwandering* mod Γ. If M is a Hadamard manifold and Γ is a properly discontinuous group of isometries of M such that $\mathrm{vol}(M/\Gamma) < \infty$, then Γ satisfies the duality condition. This is an immediate consequence of the Poincaré Recurrence Theorem. The duality condition is discussed in Section 1 of Chapter III. The technical advantages of this notion are that it is invariant under passage to subgroups of finite index and to supergroups and that it behaves well under product decompositions of X.

Theorem B. *Let M be a Hadamard manifold and Γ a group of isometries of M satisfying the duality condition. Assume that the rank of M is one. Then we have:*

(i) [Ba1] Γ contains a free non-abelian subgroup;

(ii) [Ba1] the geodesic flow of M is topologically transitive mod Γ *on the unit tangent bundle SM of M;*

(iii) [Ba1] the tangent vectors to Γ-closed geodesics of rank one are dense in SM.

Moreover, if Γ is properly discontinuous and cocompact, then we have:

(iv) [Kn] the exponential growth rate of the number of Γ-closed geodesics of rank one, where we count according to the period, is positive and equal to the topological entropy of the geodesic flow;

(v) [BB1], [Bu] the geodesic flow is ergodic mod Γ *on the set $\mathcal{R}$ of unit vectors which are tangent to geodesics of rank one;*

(vi) [Ba3], [BaL1] the Dirichlet problem at $M(\infty)$ is solvable and $M(\infty)$ together with the corresponding family of harmonic measures is the Poisson boundary of M.

Assertions (i)–(iii) are proved in Section 3 of Chapter III. The version in Chapter III is actually more general than the statements above inasmuch as we consider locally compact, geodesically complete Hadamard spaces instead of Hadamard manifolds.

Assertions (iv) and (v) are not proved in these notes. The main reason is the unclear relation between proper discontinuity and cocompactness on the one hand and the duality condition on the other in the case of general Hadamard spaces.

The ergodicity of the geodesic flow in SM/Γ for compact rank one manifolds, claimed in [BB1, Bu] (and for compact surfaces of negative Euler characteristic in [Pe2]), is based on an argument in [Pe1, 2] which contains a gap, and therefore remains an open problem (even for surfaces). The main unresolved issue is whether the set $\mathcal{R}$ has full measure in SM. In case of a surface of nonpositive curvature, $\mathcal{R}$ consists of all unit vectors which are tangent to geodesics passing through a point where the Gauss curvature of the surface is negative. Observe that $\mathcal{R}$ does have full measure if the metric on M is analytic.

For M a Hadamard manifold with a cocompact group of isometries, the relation of the rank one condition to the *Anosov condition* can be expressed in the following way:

(1) M has rank one if M has a unit speed geodesic γ such that $\dim \mathcal{J}^p(\gamma) = 1$;
(2) the geodesic flow of M is of Anosov type if $\dim \mathcal{J}^p(\gamma) = 1$ for every unit speed geodesic γ of M.

See [Eb4, 5] for this and other conditions implying the Anosov conditions for the geodesic flow.

The proofs of the various assertions in (vi) in [Ba5] and [BaL1] use a discretization procedure of Lyons and Sullivan (see [LySu, Anc3, BaL2, Kai2]) to reduce the claims to corresponding claims about random walks on Γ. In Section 4 of Chapter III we show that the Dirichlet problem at $M(\infty)$ is solvable for harmonic functions on Γ if Γ is countable and satisfies the duality condition. We also discuss briefly random walks and Poisson boundary of Γ.

Theorem B summarizes more or less what is known about rank one manifolds (as opposed to more special cases). Rank one manifolds behave in many ways like manifolds of negative sectional curvature. Rank rigidity shows that higher rank is an exceptional case.

Theorem C (Rank Rigidity). *Let M be a Hadamard manifold and assume that the group of isometries of M satisfies the duality condition. Let N be an irreducible factor in the deRham decomposition of M with* $\operatorname{rank} N = k \geq 2$. *Then N is a symmetric space of noncompact type and rank k. In particular, either M is of rank one or else a Riemannian product or a symmetric space.*

The breakthrough in the direction of this theorem was obtained in the papers [BBE] and [BBS]. Under the stronger assumptions that M admits a properly discontinuous group Γ of isometries with $\operatorname{vol} M/\Gamma < \infty$ and that the sectional curvature of M has a uniform lower bound, Theorem C was proved in [Ba2] and, somewhat later by a different method, in [BuSp]. The result as stated was proved by Eberlein (and is published in [EbHe]). The proofs in [Ba2], [BuSp] and [EbHe] build on the results in [BBE, BBS] and other papers and it may be somewhat difficult for a beginner to collect all the arguments. For that reason it seems useful to present a streamlined and complete proof in one piece. This is achieved (I hope) in Chapter IV, where the details of the proof are given. That is, we show that the holonomy group of M is not transitive on the unit sphere (at some point of M), and then the Holonomy Theorem of Berger and Simons applies, see [Be, Si].

Theorems B and C, when combined with other results, have some striking applications. We discuss some of them.

Theorem D. *Let M be a Hadamard manifold and Γ a properly discontinuous group of isometries of M with* $\operatorname{vol}(M/\Gamma) < \infty$. *Then either Γ contains a free non-abelian subgroup or else M is compact and flat and Γ a Bieberbach group.*

This follows from Theorem A if the deRham decomposition of M contains a factor of rank one (since the duality condition is preserved under products, see Section 1 in Chapter III). If there is no such factor, then M is a symmetric space by Theorem C. In this case, Γ is a linear group and the Free Subgroup Theorem of Tits applies, see [Ti1].

The following application is proved in [BaEb].

Theorem E. *Let M be a Hadamard manifold and let Γ be an irreducible properly discontinuous group of isometries of M such that* $\mathrm{vol}(M/\Gamma) < \infty$. *Then the property that M is a symmetric space of noncompact type and rank $k \geq 2$ depends only on Γ.*

The result in [BaEb] is more precise in the sense that explicit conditions on Γ are given which are equivalent to M being a symmetric space of rank $k \geq 2$. The proof in [BaEb] relies on results of Prasad-Raghunathan [PrRa] and Eberlein [Eb13] and will not be discussed here.

For the sake of completeness we mention the following immediate consequence of Theorem E and the Strong Rigidity Theorem of Mostow and Margulis [Mos, Mar1, 2].

Theorem F. *Let M, M^* be Hadamard manifolds and Γ, Γ^* be properly discontinuous groups of isometries of M and M^* respectively such that* $\mathrm{vol}(M/\Gamma)$, $\mathrm{vol}(M^*/\Gamma^*) < \infty$. *Assume that M^* is a symmetric space of noncompact type and rank $k \geq 2$, that Γ^* is irreducible and that Γ is isomorphic to Γ^*. Then M is a symmetric space of noncompact type and, up to scaling factors, isometric to M^*.*

In the cocompact case this was proved earlier by Gromov [BGS] and (under the additional assumption that M^* is reducible) Eberlein [Eb12]. Today, Theorem F does not represent the state of the art anymore, at least in the cocompact case, where superrigidity has been established in the geometric setting. This development was initiated by the striking work of Corlette [Co1, 2], and the furthest reaching results can be found in [MSY]. Another recent and exciting development is the proof of the non-equivariant Strong Rigidity Theorem by Kleiner and Leeb [KlLe].

Theorem C has the following important companion, proved by Heber [Heb].

Theorem G (Rank Rigidity for homogeneous spaces). *Let M be an irreducible homogeneous Hadamard manifold. If the rank of M is at least 2, then M is a symmetric space of noncompact type.*

Note that the isometry group of a homogeneous Hadamard manifold M satisfies the duality condition if and only if M is a symmetric space, see [Eb10, Wot1,2]. In this sense, the assumptions of Theorems C and G are complementary. In both cases, the limit set of the group of isometries is equal to $M(\infty)$, and one might wonder whether the rank rigidity holds under this weaker assumption. In fact, it might even hold without any assumption on the group of isometries.

Another interesting problem is the question of extendability of Theorems B and C to metric spaces of nonpositive Alexandrov curvature. As mentioned above already, part of that is achieved for the results in Theorem B. However, we use the flat half plane condition as in [Ba1,2] because there is, so far, no good definition of rank in the singular case. To state some of the main problems involved, assume that X is a geodesically complete and locally compact Hadamard space.

Problem 1. Assume that the isometry group of X satisfies the duality condition. Define the rank of X in such a way that
(i) $\operatorname{rank} X = k \geq 2$ if and only if every geodesic of X is contained in a k-flat;
(ii) $\operatorname{rank} X = 1$ implies some (non-uniform) hyperbolicity of the geodesic flow.

Problem 2. Assume that X is irreducible and that the group of isometries of X satisfies the duality condition. Show that X is a symmetric space or a Euclidean building if every geodesic of X is contained in a k-flat, $k \geq 2$.

Problem 3. Assume that Γ is a cocompact and properly discontinuous group of isometries of X. Show that Γ satisfies the duality condition.

Problems 1–3 have been resolved in the case of two dimensional simplicial complexes with piecewise smooth metric and cocompact group of automorphisms, see [BB3]. Kleiner has solved Problem 2 in the case $\operatorname{rank} X = \dim X$, cf. [Kl] (see also [BB3, Bar] for the case $\dim X = 2$).

Recalling the possible definitions of rank in the case of symmetric spaces we get a different notion of rank,

$$\operatorname{Rank} X \;=\; \max\{k \mid X \text{ contains a } k\text{-flat}\}$$

Clearly

$$\operatorname{rank} X \;\leq\; \operatorname{Rank} X \;\leq\; \dim X\,.$$

If X is locally compact and the isometry group of X cocompact, then the following assertions are equivalent: (i) X satisfies the Visibility Axiom of Eberlein-O'Neill; (ii) X is hyperbolic in the sense of Gromov; (iii) $\operatorname{Rank} X = 1$, see [Eb1, Gr5]. If X admits a properly discontinuous and cocompact group of isometries Γ, then

$$\operatorname{Rank} X \;\geq\; \max\{k \mid \Gamma \text{ contains a free abelian subgroup of rank } k\}$$

by the Flat Torus Theorem of Gromoll-Wolf and Lawson-Yau, see [ChEb, GroW, LaYa, Bri2]. On the other hand, Bangert and Schroeder proved that $\operatorname{Rank} M$ is the maximal integer k such Γ contains a free abelian subgroup of rank k if M is an analytic Hadamard manifold and Γ a properly discontinuous and cocompact group of isometries of M, see [BanS]. It is rather unclear whether this result can be extended to the smooth or the singular case. For more on Rank we refer to the discussion in [Gr6].

In these Lecture Notes, only topics close to the area of research of the author are discussed. References to other topics can be found in the Bibliography. It would be desirable to have complete lists of publications on the subjects of singular spaces and nonpositive curvature. The Bibliography here is a first and still incomplete step in this direction. For obvious reasons, papers about spaces of negative curvature are not included unless they are closely related to topics discussed in the text.

I am very greatful to M. Brin for contributing the Appendix on the ergodicity of geodesic flows and drawing the figures. The assistance and criticism of several people, among them the participants of the DMV-seminar in Blaubeuren and S. Alexander, R. Bishop, M. Brin, S. Buyalo, U. Lang, B. Leeb, J. Previte and P. Thurston, was very helpful when preparing the text. I thank them very much. I am also grateful to Ms. G. Goetz, who TEXed the original manuscript and its various revisions. The work on these notes was supported by the SFB256 at the University of Bonn. A good part of the manuscript was written during a stay of the author at the IHES in Bures-sur-Yvette in the fall of '94.

CHAPTER I

On the Interior Geometry of Metric Spaces

We discuss some aspects of the interior geometry of a metric space X. The metric on X is denoted d, the open respectively closed metric ball about a point $x \in X$ is denoted $B(x,r)$ or $B_r(x)$ respectively $\overline{B}(x,r)$ or $\overline{B}_r(x)$.

Good references for this chapter are [Ri, AlBe, AlBN, Gr4, BrHa].

1. Preliminaries

A *curve* in X is a continuous map $\sigma : I \to X$, where I is some interval. The *length* $L(\sigma)$ of a curve $\sigma : [a,b] \to X$ is defined as

$$L(\sigma) = \sup \sum_{i=1}^{k} d(\sigma(t_{i-1}), \sigma(t_i)) \tag{1.1}$$

where the supremum is taken over all subdivisions

$$a = t_0 < t_1 < \cdots < t_k = b$$

of $[a,b]$. If $\varphi : [a',b'] \to [a,b]$ is a homeomorphism, then $L(\sigma \circ \varphi) = L(\sigma)$. We say that σ is *rectifiable* if $L(\sigma) < \infty$. If $\sigma : [a,b] \to X$ is a rectifiable curve, then $\sigma|[t,t']$ is rectifiable for any subinterval $[t,t'] \subset [a,b]$. The *arc length*

$$s : [a,b] \to [0, L(\sigma)], \quad s(t) = L(\sigma \mid [a,t])$$

is a non-decreasing continuous surjective map and

$$\tilde{\sigma} : [0, L(\sigma)] \to X, \quad \tilde{\sigma}(s(t)) := \sigma(t)$$

is well defined, continuous and of *unit speed*, that is,

$$L(\tilde{\sigma}|[s,s']) = |s - s'| \, .$$

More generally, we say that a curve $\sigma : I \to X$ has *speed* $v \geq 0$ if

$$L(\sigma|[t,t']) = v|t - t'|$$

for all $t, t' \in I$. A curve $\sigma : I \to X$ is called a *geodesic* if σ has constant speed $v \geq 0$ and if any $t \in I$ has a neighborhood U in I such that

$$d(\sigma(t'), \sigma(t'')) = v|t' - t''| \tag{1.2}$$

for all t', t'' in U. We say that a geodesic $c : I \to X$ is *minimizing* if (1.2) holds for all $t', t'' \in I$.

The *interior metric* d_i associated to d is given by

$$d_i(x,y) := \inf\{L(\sigma) \mid \sigma \text{ is a curve from } x \text{ to } y\}. \tag{1.3}$$

One can show that d_i is a metric, except that d_i may assume the value ∞ at some pairs of points $x, y \in X$. We have

$$(d_i)_i = d_i.$$

We say that X is an *interior metric space* if $d = d_i$. An interior metric space is pathwise connected. It is easy to see that the usual distance function on a connected Riemannian manifold is an interior metric. More generally we have the following result.

1.4 Proposition. *If X is complete and for any pair x, y of points in X and $\varepsilon > 0$ there is a $z \in X$ such that*

$$d(x,z), d(y,z) \le \frac{1}{2}d(x,y) + \varepsilon\,,$$

then X is interior.

The proof of Proposition 1.4 is elementary, see [Gr4, p.4]. We omit the proof since we will not need Proposition 1.4.

We say that X is a *geodesic space* if for any pair x, y of points in X there is a minimizing geodesic from x to y.

1.5 Proposition. *If X is complete and any pair x, y of points in X has a midpoint, that is, a point $m \in X$ such that*

$$d(x,m) = d(y,m) = \frac{1}{2}d(x,y)\,,$$

then X is geodesic. More generally, if there is a constant $R > 0$ such that any pair of points $x, y \in X$ with $d(x,y) \le R$ has a midpoint, then any such pair can be connected by a minimizing geodesic.

Proof. Given x, y in X with $d(x,y) \le R$, we define a geodesic $\sigma : [0,1] \to X$ from x to y of length $d(x,y)$ in the following way: choose a midpoint m between x and y and set $\sigma(1/2) = m$. Now $d(x,m)$, $d(y,m) \le R$ and hence there are midpoints m_1 and m_2 between x and $\sigma(1/2)$ respectively $\sigma(1/2)$ and y. Set $\sigma(1/4) = m_1$, $\sigma(3/4) = m_2$. Proceeding in this way we obtain, by recursion, a map σ from the dyadic numbers in $[0,1]$ to X such that

$$d(\sigma(s), \sigma(t)) = |s-t|d(x,y)$$

for all dyadic s, t in $[0,1]$. Since X is complete, σ extends to a minimal geodesic from x to y as asserted. □

2. The Hopf-Rinow Theorem

We present Cohn-Vossen's generalization of the Theorem of Hopf-Rinow to locally compact interior metric spaces, see [Coh].

2.1 Lemma. *Let X be locally compact and interior. Then for any $x \in X$ there is an $r > 0$ such that*
(i) if $d(x, y) \leq r$, then there is a minimizing geodesic from x to y;
(ii) if $d(x, y) > r$, then there is a point $z \in X$ with $d(x, z) = r$ and

$$d(x, y) = r + d(z, y).$$

Proof. Since X is locally compact, there is an $r > 0$ such that $\overline{B}(x, 2r)$ is compact. For any $y \in X$ there is a sequence $\sigma_n : [0, 1] \to X$ of curves from x to y such that $L(\sigma_n) \to d(x, y)$ since X is interior. Without loss of generality we can assume that σ_n has constant speed $L(\sigma_n)$.

If $d(x, y) \leq r$, then $L(\sigma_n) \leq 2r$ for n sufficiently large. Then the image of σ_n is contained in $\overline{B}(x, 2r)$ and σ_n has Lipschitz constant $2r$. Hence the sequence (σ_n) is equicontinuous and has a convergent subsequence by the theorem of Arzela-Ascoli. The limit is a minimizing geodesic from x to y, hence (i).

If $d(x, y) > r$, then there is a point $t_n \in (0, 1)$ such that $d(\sigma_n(t_n), x) = r$. A limit z of a subsequence of $(\sigma_n(t_n))$ will satisfy (ii). $\square$

We say that a geodesic $\sigma : [0, \omega) \to X$, $0 < \omega \leq \infty$ is a *ray* if σ is minimizing and if $\lim_{t \to \omega} \sigma(t)$ does not exist. The most important step in Cohn-Vossen's argument is the following result.

2.2 Theorem. *Let X be locally compact and interior. Then for x, y in X there is either a minimizing geodesic from x to y or else a unit speed ray $\sigma : [0, \omega) \to X$ with $\sigma(0) = x$, $0 < \omega < d(x, y)$, such that the points in the image of σ are between x and y; that is, if z is in the image of σ, then*

$$d(x, z) + d(z, y) = d(x, y).$$

Proof. Let $x, y \in X$ and assume that there is no minimizing geodesic from x to y. Let r_1 be the supremum of all r such that there is a point z between $x = x_0$ and y with $d(x_0, z) = r$ and there is a minimizing geodesic from x_0 to z. Then $r_1 > 0$ by Lemma 2.1. We let x_1 be such a point with $\delta_1 = d(x_0, x_1) \geq r_1/2$ and σ_1 a minimizing unit speed geodesic from x_0 to x_1. Since there is no minimizing geodesic from x_0 to y we have $x_1 \neq y$. Since x_1 is between x_0 and y, any point z in the image of σ_1 is between x_0 and y. There is no minimizing geodesic σ from x_1 to y since otherwise the concatenation $\sigma_1 * \sigma$ would be a minimizing geodesic from $x = x_0$ to y. Hence we can apply the same procedure to x_1 and obtain $r_2 > 0, x_2$ between x_1 and y with $\delta_2 = d(x_1, x_2) \geq r_2/2$ and a minimizing unit speed geodesic σ_2 from x_1 to x_2. Since x_2 is between x_1 and y, any point in the image of $\sigma_1 * \sigma_2$

is between x_0 and y. In particular, $\sigma_1 * \sigma_2$ is a minimizing unit speed geodesic and $x_2 \neq y$. Proceeding inductively, we obtain a minimizing unit speed geodesic

$$\sigma = \sigma_1 * \sigma_2 * \sigma_3 \ldots : [0, \delta_1 + \delta_2 + \delta_3 + \ldots) \to X$$

such that all points on σ are between x and y. In particular, σ is minimizing. By the definition of r_1 we have

$$(r_1 + r_2 + \ldots)/2 \leq \omega := \delta_1 + \delta_2 + \ldots \leq r_1$$

It remains to show that σ is a ray. If this is not the case, the limit $\overline{x} := \lim_{t \to \omega} \sigma(t)$ exists. Since there is no minimizing geodesic from x to y we have $\overline{x} \neq y$. But then there is a (short) minimizing unit speed geodesic $\overline{\sigma} : [0, \overline{r}] \to X$ with $\overline{r} > 0$ such that $\overline{\sigma}(0) = \overline{x}$ and such that all the points on $\overline{\sigma}$ are between $\overline{x}$ and y. Then all points on $\sigma * \overline{\sigma}$ are between x and y and, in particular, $\overline{r} \leq r_n/2$ for all n by the definition of r_n. This contradicts $\overline{r} > 0$ and $r_n \to 0$. □

2.3 Theorem of Hopf-Rinow (local version). *Let X be locally compact and interior, and let $x \in X$ and $R > 0$. Then the following are equivalent:*

(i) any geodesic $\sigma : [0, 1) \to X$ with $\sigma(0) = x$ and $L(\sigma) < R$ can be extended to the closed interval $[0, 1]$;

(ii) any minimizing geodesic $\sigma : [0, 1) \to X$ with $\sigma(0) = x$ and $L(\sigma) < R$ can be extended to the closed interval $[0, 1]$;

(iii) $\overline{B}(x, r)$ is compact for $0 \leq r < R$.

Each of these implies that for any pair y, z of points in $B(x, R)$ with $d(x, y) + d(x, z) < R$ there is a minimizing geodesic from y to z.

Proof. The conclusions (iii) ⇒ (i) and (i) ⇒ (ii) are clear. We prove (ii) ⇒ (iii). Let $r_0 \in (0, R]$ be the supremum over all $r \in (0, R)$ such that $\overline{B}(x, r)$ is compact. We assume $r_0 < R$. Let (x_n) be a sequence in $\overline{B}(x, r_0)$. From (ii) and Theorem 2.2 we conclude that there is a minimizing geodesic $\sigma_n : [0, 1] \to X$ from x to x_n. Then $d(x, \sigma_n(t)) \leq t r_0$ for $0 \leq t \leq 1$, and by a diagonal argument we conclude that $\sigma_n \mid [0, 1)$ has a subsequence converging to a minimizing geodesic $\sigma : [0, 1) \to X$. By (ii), σ can be extended to 1 and clearly $\sigma(1)$ is the limit of the (corresponding) subsequence of $(\sigma_n(1)) = (x_n)$. Hence $\overline{B}(x, r_0)$ is compact.

Since X is locally compact, there is an $\varepsilon > 0$ such that $\overline{B}(y, \varepsilon)$ is compact for any $y \in \overline{B}(x, r_0)$. But then $\overline{B}(x, r_0 + \delta)$ is compact for $\delta > 0$ sufficiently small, a contradiction to the definition of r_0. □

2.4 Theorem of Hopf-Rinow (global version). *Let X be locally compact and interior. Then the following are equivalent:*

(i) X is complete;

(ii) any geodesic $\sigma : [0, 1) \to X$ can be extended to $[0, 1]$;

(iii) for some point $x \in X$, any minimizing geodesic $\sigma : [0, 1) \to X$ with $\sigma(0) = x$ can be extended to $[0, 1]$;

(iv) bounded subsets of X are relatively compact.
Each of these implies that X is a geodesic space, that is, for any pair x, y of points in X there is a minimizing geodesic from x to y. □

The local compactness is necessary. If X is the space consisting of two vertices x, y and a sequence of edges σ_n of length $1 + 1/n, n \geq 1$, then X (with the obvious interior metric) is a complete interior metric space; but X is not a geodesic space since $d(x, y) = 1$.

3. Spaces with curvature bounded from above

For $\kappa \in \mathbb{R}$, we denote by M^2_κ the model surface of constant curvature κ. Motivated by corresponding results in Riemannian geometry, we will define upper curvature bounds for X by comparing triangles in X with triangles in M^2_κ.

A *triangle* in X consists of three geodesic segments $\sigma_1, \sigma_2, \sigma_3$ in X, called the *edges* or *sides* of the triangle, whose endpoints match (in the usual way). If $\Delta = (\sigma_1, \sigma_2, \sigma_3)$ is a triangle in X, a triangle $\overline{\Delta} = (\overline{\sigma}_1, \overline{\sigma}_2, \overline{\sigma}_3)$ in M^2_κ is called an *Alexandrov triangle* or *comparison triangle* for Δ if $L(\overline{\sigma}_i) = L(\sigma_i), 1 \leq i \leq 3$. A comparison triangle exists and is unique (up to congruence) if the sides satisfy the triangle inequality, that is

$$L(\sigma_i) + L(\sigma_j) \ \leq \ L(\sigma_k) \tag{3.1a}$$

for any permutation (i, j, k) of $(1, 2, 3)$, and if the *perimeter*

$$P(\Delta) = L(\sigma_1) + L(\sigma_2) + L(\sigma_3) < 2\pi/\sqrt{\kappa}. \tag{3.1b}$$

Here and below we use the convention $2\pi/\sqrt{\kappa} = \infty$ for $\kappa \leq 0$.

3.2 Definition. We say that a triangle Δ has the CAT_κ-property, or simply: is CAT_κ, if its sides satisfy the inequalities (3.1a), (3.1b) and if

$$d(x, y) \leq d(\overline{x}, \overline{y})$$

for all points x, y on the edges of Δ and the corresponding points $\overline{x}, \overline{y}$ on the edges of the comparison triangle $\overline{\Delta}$ in M^2_κ.

In short, a triangle is CAT_κ if it is not too fat in *Comparison* to the *Alexandrov Triangle* in M^2_κ.

3.3 Lemma. *Let $\Delta = (\sigma_1, \sigma_2, \sigma_3)$ and $\Delta' = (\sigma'_1, \sigma'_2, \sigma'_3)$ be triangles in X and assume that $\sigma_3 = \sigma'_3$ and that $\sigma_2 * \sigma'_2$ is a geodesic. If $\Delta'' = (\sigma_1, \sigma_2 * \sigma'_2, \sigma'_1)$ has perimeter $< 2\pi/\sqrt{\kappa}$ and Δ, Δ' are CAT_κ, then Δ'' is CAT_κ.*

Moreover, if the length of $\sigma_3 = \sigma'_3$ is strictly smaller than the distance of the pair of points on the comparison triangle $\overline{\Delta''}$ in M^2_κ corresponding to the endpoints of $\sigma_3 = \sigma'_3$, then the distance of any pair x, y of points on the edges of Δ'', such *that x, y are not on the same edge of Δ'', is strictly smaller than the distance of the corresponding pair of points on $\overline{\Delta''}$.*

Proof. Match the comparison triangles $\overline{\Delta}$ and $\overline{\Delta}'$ in M^2_κ along the sides $\overline{\sigma}_3$ and $\overline{\sigma}'_3$. Let x be the vertex of $\sigma_3 = \sigma'_3$ opposite to σ_1 and σ'_1 respectively. Since $\sigma_2 * \sigma'_2$ is a geodesic, we have

$$d(y, y') = d(y, x) + d(x, y')$$

for $y \in \sigma_2$ and $y' \in \sigma'_2$ sufficiently close to x. We claim that the interior angle of $\overline{\Delta} \cup \overline{\Delta}'$ at $\overline{x}$ is at least π. If not, there is a point z on $\sigma_3 = \sigma'_3$ different from x such that the minimizing geodesic (in M^2_κ) from $\overline{y}$ to $\overline{y}'$ passes through $\overline{z}$. But then

$$\begin{aligned} d(y, y') = d(y, x) + d(x, y') &= d(\overline{y}, \overline{x}) + d(\overline{x}, \overline{y}') \\ > d(\overline{y}, \overline{z}) + d(\overline{z}, \overline{y}') &\geq d(y, z) + d(z, y') \geq d(y, y'), \end{aligned}$$

where we use, in the second line, that Δ and Δ' are CAT_κ. Hence we arrive at a contradiction and the interior angle at $\overline{x}$ is at least π. In particular, the sides of Δ'' satisfy the triangle inequality (3.1a).

If the angle is equal to π, then $\overline{\Delta} \cup \overline{\Delta}'$ is the comparison triangle of Δ''. In this case it is clear that Δ'' is CAT_κ. If the angle is strictly bigger than π, the comparison triangle is obtained by straightening the broken geodesic $\overline{\sigma}_2 * \overline{\sigma}'_2$ of $\overline{\Delta} \cup \overline{\Delta}'$, keeping the length of $\overline{\sigma}_1, \overline{\sigma}'_1, \overline{\sigma}_2$ and $\overline{\sigma}'_2$ fixed. Now let Q be the union of the two triangular surfaces bounded by $\overline{\Delta}$ and $\overline{\Delta}'$ and denote by Q_t the corresponding surface obtained during the process of straightening at time t. The interior distance d_t of two points on the boundary of Q_t, that is, the distance determined by taking the infimum of the lengths of curves in Q_t connecting the given points, is strictly smaller than the distance d_s of the corresponding points at a later time s. The proof of this assertion is an exercise in the trigonometry of M^2_κ. At the final time t_f of the deformation, Q_{t_f} is a triangle in M^2_κ of perimeter $< 2\pi/\sqrt{\kappa}$ and, therefore, the interior distance d_{t_f} is equal to the distance in M^2_κ. Hence the lemma follows. □

3.4 Lemma. *Let $x \in X$ and assume that any two points $y, z \in B_r(x)$ can be connected by a minimizing geodesic in X. If $r < \pi/2\sqrt{\kappa}$ and if all triangles in $B_{2r}(x)$ with minimizing sides and of perimeter $< 2\pi/\sqrt{\kappa}$ are CAT_κ, then $B_r(x)$ is convex; more precisely, for all $y, z \in B_r(x)$ there is a unique geodesic $\sigma_{yz} : [0,1] \to B_r(x)$ from y to z and σ_{yz} is minimizing and depends continuously on y and z.*

Proof. We prove first that any geodesic $\sigma : [0,1] \to B_r(x)$ is minimizing. Otherwise there would exist, by the definition of geodesics, a maximal $t_\sigma \in (0,1)$ such that $\sigma_1 = \sigma|[0, t_\sigma]$ is minimizing and a $\delta > 0$ with $\delta < t_\sigma, 1 - t_\sigma$ such that $\sigma|[t_\sigma - \delta, t_\sigma + \delta]$ is minimizing. Let $\sigma_2 = \sigma|[t_\sigma, t_\sigma + \delta]$, $y = \sigma(t_\sigma)$, $z_1 = \sigma(t_\sigma - \delta)$ and $z_2 = \sigma(t_\sigma + \delta)$. We have

$$(*) \qquad d(z_1, z_2) = d(z_1, y) + d(y, z_2).$$

On the other hand, $\sigma|[0, t_\sigma + \delta]$ is not minimizing by the definition of t_σ. Let σ_3 be a minimizing geodesic from $\sigma(0)$ to $z_2 = \sigma(t_\sigma + \delta)$. Then

$$(**) \qquad L(\sigma_3) < L(\sigma_1) + L(\sigma_2)$$

and σ_3 is contained in $B_{2r}(x)$ since $d(x, \sigma(0)) + d(x, z_2) < 2r$. Now the triangle $\Delta = (\sigma_1, \sigma_2, \sigma_3)$ has minimizing sides and is contained in $B_{2r}(x)$. In the comparison triangle $\overline{\Delta}$ of Δ we get from $(**)$ and $(*)$,

$$\begin{aligned} d(\overline{z}_1, \overline{z}_2) &< d(\overline{z}_1, \overline{y}) + d(\overline{y}, \overline{z}_2) \\ &= d(z_1, y) + d(y, z_2) \ = \ d(z_1, z_2), \end{aligned}$$

a contradiction to Δ being CAT_κ. Hence σ is minimizing.

Now let $y, z \in B_r(x)$ and σ be a minimizing geodesic from y to z. Then σ is contained in $B_{2r}(x)$ and σ together with minimizing geodesics from x to y and x to z forms a triangle Δ in $B_{2r}(x)$ with minimizing sides. Since Δ is CAT_κ and $r < \pi/2\sqrt{\kappa}$, any point on σ has distance $< r$ to x; that is, σ is in $B_r(x)$. Now Lemma 3.4 follows easily. □

3.5 Remark. The assumptions and assertions in Lemma 3.4 are not optimal; compare also Corollary 4.2 below.

3.6 Definition. For $\kappa \in \mathbb{R}$, an open subset U of X is called a *CAT_κ-domain* if and only if

(i) for all $x, y \in U$, there is a geodesic $\sigma_{xy} : [0,1] \to U$ of length $d(x,y)$;

(ii) all triangles in U are CAT_κ.

We say that X has *Alexandrov curvature at most* κ, in symbols : $K_X \le \kappa$, if every point $x \in X$ is contained in a CAT_κ-domain.

For $\kappa = 1$, the standard example of a CAT_κ-domain is the open hemisphere in the unit sphere. Euclidean space respectively (real) hyperbolic space are examples of CAT_κ-domains for $\kappa = 0$ respectively $\kappa = -1$.

If U is a CAT_κ-domain, then, by the definition of CAT_κ, triangles in U have perimeter $< 2\pi/\sqrt{\kappa}$. In particular, any geodesic in U has length $< \pi/\sqrt{\kappa}$. Otherwise we could construct (degenerate) triangles of perimeter $\ge 2\pi/\sqrt{\kappa}$. Since triangles in U are CAT_κ, we conclude that the geodesic σ_{xy} as in (i) is the unique geodesic in U from x to y and that σ_{xy} depends continuously on x and y. By the same reason, if $x \in U$ and if $B_r(x) \subset U$ for some $r < \pi/2\sqrt{\kappa}$, then $B_r(x)$ is a CAT_κ-domain.

3.7 Exercises and remarks. (a) If X is a Riemannian manifold, then the Alexandrov curvature of X is at most κ iff the sectional curvature of X is bounded from above by κ.

(b) Define a function K_X on X by

$$K_X(x) \ = \ \inf\{\kappa \in \mathbb{R} | x \text{ is contained in a } CAT_\kappa\text{-domain } U \subset X\}.$$

Show that this function is bounded from above by κ if the Alexandrov curvature of X is at most κ. (This is a reconciliation with the notation.)

(c) Let X be a geodesic space. Show that X has Alexandrov curvature at most κ iff every point x in X has a neighborhood U such that for any triangle in X with vertices in U and minimizing sides the distance of the vertices to the midpoints of the opposite sides of the triangle is bounded from above by the distance of the corresponding points in the comparison triangle in M^2_κ.

In [Ri], the interested reader finds a thorough discussion of the above and other definitions of upper curvature bounds.

Now assume that $U \subset X$ is a CAT_κ-domain. Let $\varepsilon > 0$ and let $\sigma_1, \sigma_2 : [0, \varepsilon] \to U$ be two unit speed geodesics with $\sigma_1(0) = \sigma_2(0) =: x$. For $s, t \in (0, \varepsilon)$ let Δ_{st} be the triangle spanned by $\sigma_1|[0,s]$ and $\sigma_2|[0,t]$. Let $\gamma(s,t)$ be the angle at $\overline{x}$ of the comparison triangle $\overline{\Delta}_{st}$ in M^2_κ. Then $\gamma(s,t)$ is monotonically decreasing as s, t decrease and hence

$$\angle(\sigma_1, \sigma_2) := \lim_{s,t\to 0} \gamma(s,t) \tag{3.8}$$

exists and is called the *angle* subtended by σ_1 and σ_2. The angle function satisfies the *triangle inequality*

$$\angle(\sigma_1, \sigma_3) \;\le\; \angle(\sigma_1, \sigma_2) + \angle(\sigma_2, \sigma_3)\,. \tag{3.9}$$

The triangle inequality is very useful in combination with the fact that

$$\angle(\sigma_1, \sigma_2) \;=\; \pi \quad \text{if } \sigma_1^{-1} * \sigma_2 \text{ is a geodesic}\,. \tag{3.10}$$

Here σ_1^{-1} is defined by $\sigma_1^{-1}(t) = \sigma_1(-t)$, $-\varepsilon \le t \le 0$. The trigonometric formulas for spaces of constant curvature show that we can use comparison triangles in M^2_λ as well, where $\lambda \in \mathbb{R}$ is arbitrary (but fixed), and obtain the same angle measure. In particular, we have the following formula:

$$\cos(\angle(\sigma_1, \sigma_2)) = \lim_{s,t\to 0} \frac{d^2(\sigma_1(s), \sigma_2(t)) - s^2 - t^2}{2st}\,. \tag{3.11}$$

If $y, z \in U \setminus \{x\}$ and σ_1, σ_2 are the unit speed geodesics from x to y and z respectively, we set $\angle_x(y,z) = \angle(\sigma_1, \sigma_2)$.

3.12 Exercise. In U as above let $x_n \to x$, $y_n \to y$, $z_n \to z$ with $x \in U$ and $y, z \in U \setminus \{x\}$. Then $\angle_{x_n}(y_n, z_n)$ and $\angle_x(y_n, z_n)$ are defined for n sufficiently large and

$$\angle_x(y,z) = \lim \angle_x(y_n, z_n) \ge \limsup \angle_{x_n}(y_n, z_n).$$

In a CAT_κ-domain U, we can also speak of the triangle $\Delta(x_1, x_2, x_3)$ spanned by three points x_1, x_2, x_3 since there are unique geodesic connections between these points.

3.13 Proposition. *Let U be a CAT_κ-domain, let $\Delta = \Delta(x_1, x_2, x_3)$ be a triangle in U, and let $\overline{\Delta}$ be the comparison triangle in M^2_κ.*

(i) Let α_i be the angle of Δ at $x_i, 1 \leq i \leq 3$. Then $\alpha_i \leq \overline{\alpha}_i$, where $\overline{\alpha}_i$ is the corresponding angle in $\overline{\Delta}$.

(ii) If $d(x,y) = d(\overline{x}, \overline{y})$ for one pair of points on the boundary of Δ such that x, y do not lie on the same edge, or if $\alpha_i = \overline{\alpha}_i$ for one i, then Δ bounds a convex region in U isometric to the triangular region in M^2_κ bounded by $\overline{\Delta}$.

Proof. The first assertion follows immediately from the definition of angles since triangles in U are CAT_κ. The equality in (ii) follows easily from the last assertion in Lemma 3.3. □

4. The Hadamard-Cartan Theorem

In this section, we present the proof of the Hadamard-Cartan Theorem for simply connected, complete metric spaces with non-positive Alexandrov curvature. Our version of the theorem includes the ones given in [AB1] and [Ba5]; it was motivated by a discussion with Bruce Kleiner. We start with a general result of Alexander and Bishop [AB1] about geodesics in spaces with upper curvature bounds.

4.1 Theorem. *If X is a complete metric space with $K_X \leq \kappa$, and if $\sigma : [0,1] \to X$ is a geodesic segment of length $< \pi/\sqrt{\kappa}$, then σ does not have conjugate points in the following sense: for any point x close enough to $\sigma(0)$ and any point y close enough to $\sigma(1)$ there is a unique geodesic $\sigma_{xy} : [0,1] \to X$ from x to y close to σ. Moreover, any triangle $(\sigma_{xy}, \sigma_{xz}, \sigma_{yz})$, where x is close enough to $\sigma(0)$, y and z are close enough to $\sigma(1)$ and σ_{yz} is minimizing from y to z, is CAT_κ.*

Proof. For $0 \leq L \leq L(\sigma)$ consider the following assertion $A(L)$:

Given $\varepsilon > 0$ small there is $\delta > 0$ such that for any subsegment $\overline{\sigma} = \overline{x_0 y_0}$ of σ of length at most L and any two points x, y with $d(x, x_0), d(y, y_0) < \delta$ there is a unique geodesic σ_{xy} from x to y whose distance from $\overline{\sigma}$ is less than ε. Moreover, $|L(\sigma_{xy}) - L(\overline{\sigma})| \leq \varepsilon$ and any triangle $(\sigma_{xy}, \sigma_{xz}, \sigma_{yz})$, where $d(x, x_0)$, $d(y, y_0)$, $d(z, y_0) < \delta$ and where σ_{yz} is minimizing from y to z, is CAT_κ.

Let $r > 0$ be a uniform radius such that $B_r(z)$ is a CAT_κ-domain for any point z on σ. Then $A(L)$ holds for $L \leq r$. We show now that $A(3L/2)$ holds if $A(L)$ holds and $3L/2 \leq L(\sigma) < \pi/\sqrt{\kappa}$.

To that end choose $\alpha > 0$ with $L(\sigma) + 3\alpha < \pi/\sqrt{\kappa}$. Then $L + 2\alpha < 2\pi/3\sqrt{\kappa}$ and therefore there is a constant $\lambda < 1$ such that for any two geodesics $\overline{\sigma}_1, \overline{\sigma}_2 : [0,1] \to M^2_\kappa$ with $\overline{\sigma}_1(0) = \overline{\sigma}_2(0)$ and with length at most $L + 2\alpha$ we have

$$(*) \qquad d(\overline{\sigma}_1(t), \overline{\sigma}_2(t)) \leq \lambda d(\overline{\sigma}_1(1), \overline{\sigma}_2(1)), \ 0 \leq t \leq \frac{1}{2}.$$

Now let $\overline{\sigma} = \overline{x_0 y_0}$ be a subsegment of σ of length $3l/2 \leq 3L/2$ and let $\varepsilon > 0$ be given. Choose $\varepsilon' > 0$ with $\varepsilon' < \min\{\varepsilon/3, \alpha\}$, and let $\delta < \delta'(1-\lambda)/\lambda$, where δ' is the value guaranteed by $A(L)$ for ε'.

Subdivide $\overline{\sigma}$ into thirds by points p_0 and q_0. Let x, y be points such that $d(x, x_0), d(y, y_0) < \delta$. By $A(L)$ there are unique geodesics σ_{xq_0} from x to q_0 and σ_{p_0y} from p_0 to y of distance at most ε' to $\overline{x_0q_0}$ respectively $\overline{p_0y_0}$. Furthermore, their length is in $[l-\varepsilon', l+\varepsilon']$ and the triangles $(\overline{x_0q_0}, \sigma_{xq_0}, \sigma_{xx_0})$ and $(\overline{p_0y_0}, \sigma_{p_0y}, \sigma_{yy_0})$ are CAT_κ. Hence we can apply $(*)$ to the midpoints p_1 of σ_{xq_0} and q_1 of σ_{p_0y} and obtain

$$d(p_0, p_1), d(q_0, q_1) \leq \lambda \max\{d(x, x_0), d(y, y_0)\} < \lambda\delta < \delta'.$$

By $A(L)$ there are unique geodesics σ_{xq_1} from x to q_1 and σ_{p_1y} from p_1 to y of distance at most ε' to $\overline{x_0q_0}$ respectively $\overline{p_0y_0}$. Furthermore, their length is in $[l-\varepsilon', l+\varepsilon']$ and the triangles $(\sigma_{xq_0}, \sigma_{xq_1}, \sigma_{q_0q_1})$ and $(\sigma_{p_0y}, \sigma_{p_1y}, \sigma_{p_0p_1})$ are CAT_κ. Hence we can apply $(*)$ to the midpoints p_2 of σ_{xq_1} and q_2 of σ_{p_1y} and obtain

$$d(p_1, p_2), d(q_1, q_2) \leq \lambda \max\{d(q_0, q_1), d(p_0, p_1)\} < \lambda^2\delta.$$

Hence by the triangle inequality

$$d(p_0, p_2), d(q_0, q_2) < (\lambda + \lambda^2)\delta < \delta'.$$

Recursively we obtain geodesics σ_{xq_n} from x to the midpoint q_n of $\sigma_{p_{n-1}y}$ and σ_{p_ny} from the midpoint p_n of $\sigma_{xq_{n-1}}$ to y of distance at most ε' to $\overline{x_0q_0}$ respectively $\overline{p_0y_0}$. Their length is in $[l-\varepsilon', l+\varepsilon']$ and the triangles $(\sigma_{xq_{n-1}}, \sigma_{xq_n}, \sigma_{q_{n-1}q_n})$ and $(\sigma_{p_{n-1}y}, \sigma_{p_ny}, \sigma_{p_{n-1}p_n})$ are CAT_κ. Furthermore, we have the estimates

$$d(p_{n-1}, p_n), d(q_{n-1}, q_n) < \lambda^n\delta$$

and

$$d(p_0, p_n), d(q_0, q_n) < (\lambda + \ldots + \lambda^n)\delta < \delta'.$$

In particular, the sequences (p_n), (q_n) are Cauchy. Since X is complete, they converge. If $p = \lim p_n$ and $q = \lim q_n$, then

$$d(p_0, p), d(q_0, q) \leq \frac{\lambda}{1-\lambda}\delta < \delta'$$

and hence, by $A(L)$, the geodesics σ_{xq_n} and σ_{p_ny} converge to σ_{xq} and σ_{py}. By construction, $p \in \sigma_{xq}$ and $q \in \sigma_{py}$. Therefore, by the uniqueness of σ_{pq}, the geodesics σ_{xq} and σ_{py} overlap in σ_{pq} and combine to give a geodesic σ_{xy} from x to y. The length of σ_{xy} is given by $L(\sigma_{xq}) + L(\sigma_{py}) - L(\sigma_{pq})$, hence $|L(\sigma_{xy}) - L(\sigma)| < \varepsilon$ by $A(L)$ and since $\varepsilon' < \varepsilon/3$. The last assertion follows from Lemma 3.3 by subdividing the triangle suitably. □

4.2 Lemma. *Let X be a complete metric space with $K_X \leq \kappa$, and let $\sigma_0 : [0,1] \to X$ be a geodesic segment of length $L_0 = L(\sigma_0) < \pi/\sqrt{\kappa}$. For $\varepsilon > 0$ with $L + 2\varepsilon < \pi/\sqrt{\kappa}$ assume that the balls $B_\varepsilon(x_0)$ and $B_\varepsilon(y_0)$ are CAT_κ-domains, where $x_0 = \sigma_0(0)$ and $y_0 = \sigma_0(1)$. Then there is a continuous map*

$$\Sigma : B_\varepsilon(x_0) \times B_\varepsilon(y_0) \times [0,1] \to X$$

with the following properties:

(i) $\sigma_{xy} = \Sigma(x, y, .)$ *is a geodesic from* x *to* y *of length* $L(\sigma_{xy}) \leq L_0 + d(x, x_0) + d(y, y_0)$ *and* $\sigma_{x_0 y_0} = \sigma_0$;

(ii) any triangle $(\sigma_{xy}, \sigma_{xz}, \sigma)$, *where* σ *is the geodesic in* $B_\varepsilon(y_0)$ *from* y *to* z, *is* CAT_κ.

Moreover, Σ *is unique in the following sense: if* $\sigma_s, 0 \leq s \leq 1$, *is a homotopy of* σ_0 *by geodesics with* $\sigma_s(0) = x_s \in B_\varepsilon(x_0)$ *and* $\sigma_s(1) = y_s \in B_\varepsilon(x_1)$, *then* $\sigma_s = \sigma_{x_s y_s}$.

Proof. The assertion about the uniqueness is immediate from the uniqueness assertion in Theorem 4.1. Now let $x \in B_\varepsilon(x_0)$, $y \in B_\varepsilon(y_0)$ and let

$$\gamma_0 : [0, 1] \to B_\varepsilon(x_0), \quad \gamma_1 : [0, 1] \to B_\varepsilon(y_0)$$

be the geodesics from x_0 to x respectively y_0 to y. Let $r_0 \geq 0$ be the supremum over all $r \in [0, 1]$ such that there is a continuous map

$$F_r : [0, r] \times [0, r] \times [0, 1] \to X$$

with $F_r(0, 0, .) = \sigma_0$ and

(i') $\sigma_{st} = F_r(s, t, .)$ is a geodesic from $\gamma_0(s)$ to $\gamma_1(t)$ of length $L(\sigma_{st}) \leq L_0 + d(x_0, \gamma_0(s)) + d(y_0, \gamma_1(t))$;

(ii') triangles $(\gamma_0 \mid [s_1, s_2]), \sigma_{s_1 t}, \sigma_{s_2 t})$ and $(\sigma_{st_1}, \sigma_{st_2}, \gamma_1 \mid [t_1, t_2])$ are CAT_κ.

The uniqueness assertion in Theorem 4.1 implies that F_{r_1} and F_{r_2} agree on their common domain of definition for all $r_1, r_2 \in [0, r_0)$. Now $L + 2\varepsilon < \pi/\sqrt{\kappa}$, and hence the circumference of triangles as in (ii'), for s_1, s_2 respectively t_1, t_2 small, is uniformly bounded away from $2\pi/\sqrt{\kappa}$. Comparison with triangles in M^2_κ shows that $\lim_{r \to r_0} F_r =: F_{r_0}$ exists. It is clear that $F_{r_0}(0, 0, .) = \sigma_0$ and that F_{r_0} satisfies (i') and (ii'). This shows that the set J of r, for which a map F_r as above exists, is closed in $[0, 1]$. On the other hand, J is open by Theorem 4.1. Hence $r_0 = 1$.

We set $F_{xy} = F_1$ and $\sigma_{xy} = F_1(1, 1, .)$. It follows from Theorem 4.1 and uniqueness that F_{xy}, and hence σ_{xy}, depends continuously on x and y. Assertion (ii) follows from Lemma 3.3. □

4.3 Corollary. *Let* X *be a complete interior metric space with* $K_X \leq \kappa$. *Let* $R \leq \pi/\sqrt{\kappa}$ *and assume that for any two points* x, y *with* $d(x, y) < R$ *there is a unique geodesic* $\sigma_{xy} : [0, 1] \to X$ *from* x *to* y *of length* $< R$. *Then*

(i) $L(\sigma_{xy}) = d(x, y)$ *for all such* x, y;

(ii) each triangle in X *of perimeter* $< 2R$ *is* CAT_κ.

In particular, $B_{R/2}(x)$ *is a* CAT_κ*-domain for any* $x \in X$.

Proof. Since X is interior, there is a curve $c : [0, 1] \to X$ from x to y of length $< R$. By Lemma 4.2, such a curve c gives rise to a homotopy σ_s, $0 \leq s \leq 1$, such that $\sigma_s(t) = c(t)$ for $s \leq t \leq 1$ and $\sigma_s(t)$, $0 \leq t \leq s$, is a geodesic from x to $c(s)$ which is not longer than $c(t)$, $0 \leq t \leq s$. In particular, σ_1 is a geodesic from x to y of length $\leq L(c) < R$. Hence $\sigma_1 = \sigma_{xy}$, independently of the choice of c. Since

X is interior we obtain $L(\sigma_{xy}) = d(x, y)$, hence (i). The remaining assertions of Corollary 4.3 follow easily. $\square$

We now apply the above argument in the case of a family of curves. To avoid asumptions on the lengths of the curves, we assume that the Alexandrov curvature is nonpositive.

4.4 Lemma. *Let X be a complete metric space with $K_X \leq 0$. Let $f : K \times I \to X$ be a continuous map, where K is a compact space and $I = [0,1]$. Then there is a homotopy $F : I \times K \times I \to X$ of f such that*

(i) $F(s,k,t) = f(k,t)$ for all $k \in K$ and $t \geq s$;
(ii) $F(s,k,t), 0 \leq t \leq s$, is a geodesic from $f(k,0)$ to $f(k,s)$.
(iii) $F(s,k,t), 0 \leq t \leq s$, is not longer than $f(k,t), 0 \leq t \leq s$;
(iv) $d(F(s,k,t), F(s,k',t)) \leq rd(f(k,0), f(k',0)) + (1-r)d(f(k,s), f(k',s))$, $r = t/s$, if k and k' are sufficiently close.

Proof. The existence of a map F satisfying (i) and (ii) is immediate from Lemma 4.2. Since the Alexandrov curvature of X is nonpositive, (iii) and (iv) follows from (ii) in Lemma 4.2. $\square$

4.5 Theorem of Hadamard-Cartan. *Let X be a simply connected complete metric space with $K_X \leq 0$. Then for any pair x, y of points in X there is a unique geodesic $\sigma_{xy} : [0,1] \to X$ from x to y. Moreover, σ_{xy} is continuous in x and y and $L(\sigma_{xy}) = d_i(x,y)$, where d_i denotes the interior metric associated to d.*

Proof. Let $f : [0,1] \to X$ be a path from x to y. By Lemma 4.4 there is a homotopy F of f such that $\sigma = F(1,.)$ is a geodesic from x to y. This proves existence. Note also that $L(\sigma) \leq L(f)$.

As for uniqueness, if σ_0 and σ_1 are geodesics from x to y, then there is a homotopy f between them fixing the endpoints x and y. Consider the associated map F as in Lemma 4.4. The uniqueness assertion in Lemma 4.2 implies $F(1,0,.) = \sigma_0$ and $F(1,1,.) = \sigma_1$. Now

$$F(1,0,0) = F(1,1,0) = x \quad \text{and} \quad F(1,0,1) = F(1,1,1) = y$$

and therefore $F(1,0,.) = F(1,1,.)$ by (iv) in Lemma 4.4. Hence $\sigma_0 = \sigma_1$.

We have proved that, for any pair x, y of points in X, there is a unique geodesic $\sigma_{xy} : [0,1] \to X$ from x to y. By Lemma 4.2, σ_{xy} is continuous in x and y. The uniqueness of $\sigma = \sigma_{xy}$ implies that $L(\sigma) \leq L(f)$, where f is any curve from x to y. Hence $L(\sigma) = d_i(x,y)$. $\square$

5. Hadamard spaces

A *Hadamard manifold* is a simply connected, complete smooth Riemannian manifold without boundary and with nonpositive sectional curvature. Following this terminology, we say that a metric space X is a *Hadamard space* if X is simply connected, complete, geodesic with $K_X \leq 0$. A Hadamard space need not be *geodesically complete*, that is, we do not require that every geodesic segment of X is contained in a geodesic which is defined on the whole real line.

The following elegant characterization of Hadamard spaces can be found in the paper of Bruhat and Tits [BruT], see also [Bro,p.155].

5.1 Proposition. *Let X be a complete metric space. Then X is a Hadamard space if and only if for any pair x, y of points in X there is a point $m \in X$ (the midpoint between x and y) such that*

$$d^2(z,m) \leq \frac{1}{2}(d^2(z,x) + d^2(z,y)) - \frac{1}{4}d^2(x,y) \quad \textit{for all } z \in X.$$

Using Karcher's modified distance function [Kar] one obtains a similar formula characterizing complete geodesic spaces X with $K_X \leq \kappa$ such that for any pair of points $x, y \in X$ with $d(x,y) < R$ (for some fixed $R > 0$, $R \leq \pi/\sqrt{\kappa}$) there is a unique geodesic from x to y of length $< R$.

Proof of Proposition 5.1. Substituting x and y for z we see that m is in the middle between x and y, $d(x,m) = d(y,m) = \frac{1}{2}d(x,y)$, hence X is geodesic, see Proposition 1.5. It is also immediate that m is unique and varies continuously with the endpoints, hence the geodesic connecting x and y is unique and varies continuously with x and y. Therefore X is contractible and, in particular, simply connected. Now if x, y, z are the vertices of a geodesic triangle in X, then the inequality in Proposition 5.1 asserts that the distance of z to m is bounded by the distance of $\bar{z}$ to $\bar{m}$ in the comparison triangle in the Euclidean plane. Therefore, $K_X \leq 0$, compare Exercise 3.7. □

We now collect consequences of our discussion in the previous sections, in particular of Proposition 3.13, Corollary 4.3 and the Theorem of Hadamard-Cartan. The main property is that for any pair x, y of points in a Hadamard space X, there is a unique geodesic $\sigma_{xy} : [0,1] \to X$ connecting x and y.

5.2 Proposition. *Let X be a Hadamard space and let Δ be a triangle in X with edges of length a, b and c and angles α, β and γ at the opposite vertices respectively. Then*

(i) $\alpha + \beta + \gamma \leq \pi$;

(ii) $c^2 \geq a^2 + b^2 - 2ab\cos\gamma$ *(First Cosine Inequality)*

(iii) $c \leq b\cos\alpha + a\cos\beta$ *(Second Cosine Inequality).*

In each case, equality holds if and only if Δ is flat, that is, Δ bounds a convex region in X isometric to the triangular region bounded by the comparison triangle in the flat plane. □

The following formula is useful in connection with the Second Cosine Inequality, see [BGS] or respectively the corresponding argument in the proof of Theorem II.4.3.

5.3 Proposition. *Let X be a Hadamard space and let Δ be a triangle in X with edges of length a, b and c and angles α, β and γ in the opposite vertices. Let α_E and β_E be the angle in the Euclidean triangle Δ_E spanned by two edges of length a and b which subtend an angle of measure $\gamma_E = \pi - \alpha - \beta$. Then*

$$b \cos \alpha + a \cos \beta = c_E \cos(\alpha - \alpha_E) = c_E \cos(\beta - \beta_E)$$

where c_E is the length of the third edge of Δ_E. In particular, $b \cos \alpha + a \cos \beta = c_E$ if and only if Δ is flat.

Proof. Consider the function

$$f(s) = b \cos s + a \cos(\alpha + \beta - s).$$

Then

$$f' = -b \sin s + a \sin(\alpha + \beta - s) \quad \text{and} \quad f'' = -f.$$

By the Law of Sines, f assumes its maximum c_E in α_E and hence we have $f(s) = c_E \cos(s - \alpha_E)$, where c_E is the length of the third edge in Δ_E. □

5.4 Proposition. *Let I be an interval, and let $\sigma_1, \sigma_2 : I \to X$ be two geodesics in a Hadamard space X. Then $d(\sigma_1(t), \sigma_2(t))$ is convex in t.*

Proof. Let $t_0 < t_1$ be points in I, and let $\sigma : [t_0, t_1] \to X$ be the geodesic from $\sigma_1(t_0)$ to $\sigma_2(t_1)$. Since triangles in X are CAT_0, we have, for $t = \frac{1}{2}(t_0 + t_1)$,

$$\begin{aligned} d(\sigma_1(t), \sigma_2(t)) &\leq d(\sigma_1(t), \sigma(t)) + d(\sigma(t), \sigma_2(t)) \\ &\leq \frac{1}{2} d(\sigma_1(t_1), \sigma_2(t_1)) + \frac{1}{2} d(\sigma_1(t_0), \sigma_2(t_0)). \end{aligned}$$

□

5.5 Remark. The convexity of $d(\sigma_1(t), \sigma_2(t))$ in t, for any pair of geodesics σ_1, σ_2 in X, is not equivalent to X being a Hadamard space. For example, if X is a Banach space with strictly convex unit ball, then geodesics are straight lines and hence the above convexity property holds. On the other hand, a Banach space has an upper curvature bound in the sense of Definition 3.6 if and only if it is Euclidean, see [Ri].

We say that a function f on a geodesic space X is *convex* if $f \circ \sigma$ is convex for any geodesic σ in X.

5.6 Corollary. *Let X be a Hadamard space and $C \subset X$ a convex subset. Then*

$$d_C : X \to \mathbb{R}, \ d_C(z) = d(z, C),$$

is a convex function. If C is closed, then for any $z \in X$ there is a unique point $\pi_C(z) \in C$ with $d(z, \pi_C(z)) = d_C(z)$. The map π_C is called the projection onto C; it has Lipschitz constant 1.

Proof. Everything except for the existence of the point $\pi_C(z)$ is clear. Now let (x_n) be a sequence in C with $d(z, x_n) \to d_C(z)$. Since C is convex, the midpoint

$m = m_{nl}$ between x_n and x_l is also in C. Its distance to z satisfies the estimate in Proposition 5.1, where we subsitute x_n and x_l for x and y. It follows that (x_n) is a Cauchy sequence. Now X is complete and C is closed, hence the sequence has a limit and the limit is in C. By continuity, it realizes the distance from z to C. □

5.7 Proposition. *Let X be a Hadamard space and let $\square = (\sigma_1, \sigma_2, \sigma_3, \sigma_4)$ be a quadrangle of four consecutive geodesic segments in X with interior angle α_i subtended by σ_i and σ_{i+1} at their common vertex (where we count indices mod 4). Then $\alpha_1 + \alpha_2 + \alpha_3 + \alpha_4 \le 2\pi$, and equality holds if and only if $\square$ is flat, that is, $\square$ bounds a convex region isometric to a convex region in the flat plane bounded by four line segments.*

Proof. Subdivide $\square$ into two triangles by the geodesic from the initial point of σ_1 to the endpoint of σ_2 and apply (i) of Proposition 5.2. □

We say that two unit speed rays $\sigma_1, \sigma_2 : [0, \infty) \to X$ (respectively unit speed geodesics $\sigma_1, \sigma_2 : \mathbb{R} \to X$) are *asymptotic* (respectively *parallel*) if $d(\sigma_1(t), \sigma_2(t))$ is uniformly bounded.

5.8 Corollary. *Let X be a Hadamard space.*

(i) *If $\sigma_1, \sigma_2 : [0, \infty) \to X$ are asymptotic unit speed rays, then*

$$\angle_{\sigma_1(0)}(\sigma_2(0), \sigma_1(1)) + \angle_{\sigma_2(0)}(\sigma_1(0), \sigma_2(1)) \le \pi$$

and equality holds if and only if σ_1, σ_2 and the geodesic σ from $\sigma_1(0)$ to $\sigma_2(0)$ bound a flat half strip: a convex region isometric to the convex hull of two parallel rays in the flat plane.

(ii) *(Flat Strip Theorem) If $\sigma_1, \sigma_2 : \mathbb{R} \to X$ are parallel unit speed geodesics, then σ_1 and σ_2 bound a flat strip: a convex region isometric to the convex hull of two parallel lines in the flat plane.* □

5.9. Proposition. *Let X be a Hadamard space and $\sigma : \mathbb{R} \to X$ a unit speed geodesic. Then the set $P = P_\sigma \subset X$ of all points in X which belong to geodesics parallel to σ is closed, convex and splits isometrically as $P = Q \times \mathbb{R}$, where $Q \subset X$ is closed and convex and $\{q\} \times \mathbb{R}$ is parallel to σ for any $q \in Q$.*

Proof. The convexity of P follows immediately from the Flat Strip Theorem 5.8(ii). The closedness of P follows from the completeness of X. Now set $q_0 = \sigma(0)$ and let $x \in P$. Then there is, up to parameterization, a unique unit speed geodesic σ_x parallel to σ with $x \in \sigma_x$. Denote by q_x the point on σ_x closest to q_0 and choose the parameter on σ_x such that $\sigma_x(0) = q_x$. Let $t_x \in \mathbb{R}$ be the unique parameter value with $x = \sigma_x(t_x)$. It follows from the Flat Strip Theorem that q_x (respectively q_0) is the unique point on σ_x (respectively σ) with $d(\sigma(t), q_x) - t \to 0$ (respectively $d(\sigma_x(t), q_0) - t \to 0$) as $t \to \infty$.

Now let $y \in P$ and define q_y and t_y accordingly. Then we have

$$d(\sigma_y(t), q_x) - t \le d(\sigma_y(t), \sigma(t/2)) - \frac{t}{2} + d(\sigma(t/2), q_x) - \frac{t}{2}$$

and hence, since $d(\sigma_y(t), \sigma(t/2)) - t/2 \to 0$ as $t \to \infty$,

$$(*) \qquad \limsup_{t\to\infty} d(\sigma_y(t), q_x) - t \le 0.$$

Similarly

$$(**) \qquad \limsup_{t\to\infty} d(\sigma_x(t), q_y) - t \le 0.$$

Now σ_x and σ_y are parallel and bound a flat strip. From $(*)$ and $(**)$ we conclude that q_y is the closest point to q_x on σ_y and q_x is the closest point to q_y on σ_x. Hence (don't forget the flat strip)

$$d^2(x,y) = d^2(q_x, q_y) + (t_x - t_y)^2.$$

Hence the assertion with $Q = \{q_x \mid x \in P\}$. □

Following the presentation in [Bro], we now discuss circumcenter and circumscribed balls for bounded subsets in Hadamard spaces.

5.10 Proposition. *Let X be a Hadamard space and let $A \subset X$ be a bounded subset. For $x \in X$ let $r(x,A) = \sup_{y\in A} d(x,y)$ and set $r(A) = \inf_{x\in X} r(x,A)$. Then there is a unique $x \in X$, the circumcenter of A, such that $r(x,A) = r(A)$, that is, such that $A \subset \overline{B}(x, r(A))$.*

Proof. Let $x, y \in X$ and let m be the midpoint between them. Then

$$r^2(m,A) \le \frac{1}{2}(r^2(x,A) + r^2(y,A)) - \frac{1}{4}d^2(x,y),$$

see Proposition 5.1, and hence

$$\begin{aligned} d^2(x,y) &\le 2(r^2(x,A) + r^2(y,A)) - 4r^2(m,A) \\ &\le 2(r^2(x,A) + r^2(y,A)) - 4r^2(A). \end{aligned}$$

Now the uniqueness of the circumcenter is immediate. It also follows that a sequence (x_n) with $r(x_n, A) \to r(A)$ is Cauchy. Now X is complete, hence we infer the existence of a circumcenter. □

With similar arguments one obtains centers of gravity for measures on Hadamard spaces.

5.11 Exercise. Let X be a Hadamard space and let μ be a measure on X such that $g(x) = \int d^2(x,y)\mu(dy)$ is finite for one (and hence any) $x \in X$. Show that g assumes its infimum at precisely one point. This point is called the *barycenter* or *center of gravity* of μ.

Barycenters and circumcenters can also be defined in CAT_κ-domains; compare [BuKa] for the discussion in the Riemannian case.

CHAPTER II

The Boundary at Infinity

In this chapter, we present a variation of §§ 3–4 of [BGS]. We assume throughout that X is a complete geodesic space. The various spaces of maps discussed are assumed to be endowed with the topology of uniform convergence on bounded subsets.

1. Closure of X via Busemann functions

On X we consider the space $C(X)$ of continuous functions, endowed with the topology of uniform convergence on bounded subsets. For $x, y, z \in X$ we set

$$b(x,y,z) = d(x,z) - d(x,y). \tag{1.1}$$

Then we have $b(x,y,y) = 0$ and

$$|b(x,y,z) - b(x,y,z')| \leq d(z,z'), \tag{1.2}$$

and hence $b(x,y,.) \in C(X)$. Furthermore

$$|b(x,y,z) - b(x',y,z)| \leq 2d(x,x') \tag{1.3}$$

and

$$b(x,y,z) - b(x,y,y') = b(x,y',z). \tag{1.4}$$

For $y \in X$ fixed we have the map

$$b_y : X \to C(X), \quad b_y(x) = b(x,y,.).$$

It follows from (1.3) that b_y is continuous and from (1.2) that $b_y(x)$ has Lipschitz constant 1 for all $x \in X$. If $x, x' \in X$ and, say, $d(x',y) \geq d(x,y)$, then

$$b_y(x)(x') - b_y(x')(x') \geq d(x,x'),$$

and hence b_y is injective. In fact, b_y is an embedding since $b_y(x)$ and $b_y(x')$ are strictly separated if $d(x,x')$ is large. To see this suppose $d(x',y) \geq 2d(x,y)+1$ and let $z \in \overline{B}_r(x)$ be the point on the geodesic form x to x' with $d(x,z) = r := d(x,y)+1$. Then

$$\begin{aligned} b_y(x)(z) - b_y(x')(z) &= 1 - d(x',z) + d(x',y) \\ &\geq 1 + d(x',x) - d(x,y) - d(x',z) \;=\; 2. \end{aligned}$$

Now r depends only on x and not on x'. Hence b_y is an embedding.

We say that a sequence (x_n) in X converges at infinity if $d(x_n, y) \to \infty$ and $b_y(x_n)$ converges in $C(X)$ for some $y \in X$. From (1.4) we conclude that this is independent of the choice of y. Two such sequences (x_n) and (x'_n) will be called *equivalent* if $\lim_{n\to\infty} b_y(x_n) = \lim_{n\to\infty} b_y(x'_n)$ for one and hence any $y \in X$. We denote by $X(\infty)$ the set of equivalence classes. For any $\xi \in X(\infty)$ and $y \in X$ there is a well defined function $f = b(\xi, y, .) \in C(X)$, called the *Busemann function* at ξ based at y, namely $f = \lim_{n\to\infty} b_y(x_n)$, where (x_n) represents ξ. The sublevels $f^{-1}(-\infty, a)$ of f are called *horoballs* and the levels *horospheres* (centered at ξ).

Note that, by the above argument, $X(\infty)$ corresponds to the points in $\overline{b_y(X)} \setminus b_y(X)$. Hence the embedding b_y gives a topology on $X(\infty)$ and $\overline{X} = X \cup X(\infty)$. Because of (1.4) this topology does not depend on the choice of y. With respect to this topology, the function b extends to a continuous function

$$b : \overline{X} \times X \times X \to \mathbb{R}$$

such that (1.2) and (1.4) still hold and $b(x, y, y) = 0$ for all $x \in \overline{X}$ and $y \in X$.

1.5 Remarks. (a) If X is locally compact, then $\overline{X}$ and $X(\infty)$ are compact. For this note that we can apply the Theorem of Arzela-Ascoli since $b_y(x)$ is normalized by $b_y(x)(y) = 0$ and since $b_y(x)$ has Lipschitz constant 1 for all $x \in X$.

(b) In potential theory, one uses Green's functions $G(x, y)$ to define the *Martin boundary* in an analogous way. Instead of using differences $b(x, y, z) = d(x, z) - d(x, y)$, one takes quotients $K(x, y, z) = G(x, z)/G(x, y)$.

2. Closure of X via rays

In this section, we add the assumption that X is a Hadamard space. We describe the construction of $X(\infty)$ by asymptote classes of rays, introduced by Eberlein-O'Neill [EbON].

We recall that a ray is a geodesic $\sigma : [0, \omega) \to X$, $0 < \omega \le \infty$, such that σ is minimizing and such that $\lim_{t\to\omega} \sigma(t)$ does not exist. Since we assume that X is complete, we have $\omega = \infty$ for any unit speed ray in X. As in Section I.5 we say that two unit speed rays σ_1, σ_2 are *asymptotic* if $d(\sigma_1(t), \sigma_2(t))$ is bounded uniformly in t. This is an equivalence relation on the set of unit speed rays in X; the set of equivalence classes is denoted $X(\infty)$. If $\xi \in X(\infty)$ and σ is a unit speed ray belonging to ξ, we write $\sigma(\infty) = \xi$.

Recall that $d(\sigma_1(t), \sigma_2(t))$ is convex in t. In particular, if σ_1 is asymptotic to σ_2 and $\sigma_1(0) = \sigma_2(0)$, then $\sigma_1 = \sigma_2$. Hence for any $\xi \in X(\infty)$ and any $x \in X$, there is at most one unit speed ray σ starting in x with $\sigma(\infty) = \xi$. Our first aim is to show that, in fact, for each $x \in X$ and each $\xi \in X(\infty)$ there is a unit speed ray σ starting at x with $\sigma(\infty) = \xi$.

2.1 Lemma. *Let $\sigma : [0, \infty) \to X$ be a unit speed ray and let $\xi = \sigma(\infty)$. Let $x \in X$ and for $T > 0$ let $\sigma_T : [0, d(x, \sigma(T))] \to X$ be the unit speed geodesic from x to $\sigma(T)$. Then, for $R > 0$ and $\varepsilon > 0$ given, we have, for S, T sufficiently large,*

$$d(\sigma_S(t), \sigma_T(t)) < \varepsilon, \quad 0 \le t \le R.$$

Hence σ_T converges, for $T \to \infty$, to a unit speed ray $\sigma_{x,\xi}$ asymptotic to σ. That is, for $R > 0$ and $\varepsilon > 0$ given, we have, for T sufficiently large,

$$d(\sigma_{x,\xi}(t), \sigma_T(t)) < \varepsilon, \quad 0 \leq t \leq R.$$

Proof. Let $\alpha_T = \angle_{\sigma(T)}(x, \sigma(0))$. Then $\alpha_T \to 0$ since $d(\sigma(0), \sigma(T)) = T \to \infty$. Hence, for $\alpha > 0$ given, we have $\angle_{\sigma(T)}(x, \sigma(S)) \geq \pi - \alpha$ for T large and $S > T$. Hence the assertion follows from a comparison with Euclidean geometry. □

The above lemma together with our previous considerations shows that for any $x \in X$ and any $\xi \in X(\infty)$ there is a unique unit speed ray $\sigma_{x,\xi} : [0, \infty) \to X$ with $\sigma_{x,\xi}(0) = x$ and $\sigma_{x,\xi}(\infty) = \xi$. For $y \in X$, $y \neq x$, we denote by $\sigma_{x,y} : [0, d(x,y)] \to X$ the unique unit speed geodesic from x to y. On $\overline{X} = X \cup X(\infty)$ we introduce a topology by using as a basis the open sets of X together with the sets

$$U(x, \xi, R, \varepsilon) = \{z \in \overline{X} | z \notin B(x, R), d(\sigma_{x,z}(R), \sigma_{x,\xi}(R)) < \varepsilon\},$$

where $x \in X, \xi \in X(\infty)$. Recall that, by convexity, $d(\sigma_{x,z}(R), \sigma_{x,\xi}(R)) < \varepsilon$ implies that

$$d(\sigma_{x,z}(t), \sigma_{x,\xi}(t)) < \varepsilon,\ 0 \leq t \leq R.$$

The following lemma is an immediate consequence of Lemma 2.1.

2.2 Lemma. *Let $x, y \in X$, $\xi, \eta \in X(\infty)$ and let $R > 0$, $\varepsilon > 0$ be given. Assume $\eta \in U(x, \xi, R, \varepsilon)$. Then there exist $S > 0$ and $\delta > 0$ such that*

$$U(y, \eta, S, \delta) \subset U(x, \xi, R, \varepsilon).$$

□

This lemma shows that for a fixed $x \in X$ the sets $U(x, \xi, R, \varepsilon)$ together with the open subsets of X form a basis for the topology of $\overline{X}$. With respect to this topology, a sequence (x_n) in $\overline{X}$ converges to a point $\xi \in X(\infty)$ if and only if $d(x, x_n) \to \infty$ for some (and hence any) $x \in X$ and the geodesics σ_{x,x_n} converge to $\sigma_{x,\xi}$. Note also that, for any $x \in X$, the (relative) topology on $X(\infty)$ is defined by the family of pseudometrics $d_{x,R}$, $R > 0$, where

$$d_{x,R}(\xi, \eta) = d(\sigma_{x,\xi}(R), \sigma_{x,\eta}(R)). \tag{2.3}$$

Our next aim is to show that $\overline{X}$ and $X(\infty)$ are homeomorphic to the corresponding spaces in Section 1. To that end, let $x_0 \in X$ and $R > 1$ be given and assume $x_1, x_2 \in X$ satisfy $d(x_0, x_1), d(x_0, x_2) > R$. Let $y_1 = \sigma_{x_0,x_1}(R)$, $y_2 = \sigma_{x_0,x_2}(R)$ and assume $d(y_1, y_2) \geq \varepsilon$. By comparison with the Euclidean triangle we get

$$\cos(\angle_{y_1}(y_2, x_0)) \geq \frac{\varepsilon}{2R}$$

since $d(y_1, x_0) = d(y_2, x_0) = R$. Now

$$\angle_{y_1}(y_2, x_1) \geq \pi - \angle_{y_1}(y_2, x_0)$$

since y_1 is on the geodesic from x_0 to x_1, hence

$$\cos(\angle_{y_1}(y_2, x_1)) \leq \cos(\pi - \angle_{y_1}(y_2, x_0)) \leq \frac{-\varepsilon}{2R}.$$

From the First Cosine Inequality we obtain (for $\varepsilon < 1$)

$$d^2(x_1, y_2) \geq d^2(x_1, y_1) + \varepsilon^2 + \frac{\varepsilon^2}{R} d(x_1, y_1) \geq (d(x_1, y_1) + \frac{\varepsilon^2}{2R})^2$$

and hence

$$b(x_1, x_0, y_2) - b(x_2, x_0, y_2) \geq \frac{\varepsilon^2}{2R} \tag{2.4}$$

where $b(x_i, x_0, .)$ is defined as in (1.1). From this we obtain the desired conclusion about $\overline{X}$ and $X(\infty)$.

2.5 Proposition. *Let (x_n) be a sequence in X such that $d(x_0, x_n) \to \infty$. Then $b(x_n, x_0, .)$ converges to a Busemann function f if and only if σ_{x_0,x_n} converges to a ray $\sigma = \sigma_{x_0,\xi}$. Furthermore, we have $f = b_\sigma$, where*

$$b_\sigma(x) := \lim_{t\to\infty} (d(\sigma(t), x) - t).$$

Proof. If $b(x_n, x_0, .) \to f$, then $b(x_n, x_0, .)$ converges uniformly to f on $\overline{B}(x_0, R)$ for any $R > 1$ (by the choice of topology on $C(X)$). By (2.4), the sequence $(\sigma_{x_0,x_n}(R))$ converges; hence σ_{x_0,x_n} converges to a ray σ.

For the proof of the converse, note that for $r > 0$ and $\varepsilon > 0$ given, there is an $R > 0$ such that

$$|b(y, x_0, x) - b(z, x_0, x)| < \varepsilon$$

for all $x \in \overline{B}(x_0, r)$ if $d(z, x_0) > R$ and $y = \sigma_{x_0,z}(R)$. $\square$

2.6 Exercise. There is the following intrinsic characterization of Busemann functions in the case of Hadamard spaces: a function $f : X \to \mathbb{R}$ is a Busemann function based at $x_0 \in X$ if and only if
(i) $f(x_0) = 0$; (ii) f is convex; (iii) f has Lipschitz constant 1; (iv) for any $x \in X$ and $r > 0$ there is a $z \in X$ with $d(x, z) = r$ and $f(x) - f(z) = r$.

This characterization of Busemann functions is a variation of the one given in Lemma 3.4 of [BGS]. It has been suggested that a proof of Exercise 2.6 is included. The interested reader finds it in Section 3 of Chapter IV. Exercise 2.6 will be used in the discussion of fixed points of isometries.

3. Classification of isometries

Let X be a Hadamard space and let $\varphi : X \to X$ be an isometry. The function

$$d_\varphi : X \to \mathbb{R},\ d_\varphi(x) = d(x, \varphi(x)),$$

is called the *displacement function* of φ. Since φ maps geodesics to geodesics, d_φ is convex.

Definition 3.1. We say that φ is *semisimple* if d_φ achieves its mimimum in X. If φ is semisimple and $\min d_\varphi = 0$, we say that φ is *elliptic*. If φ is semisimple and $\min d_\varphi > 0$, we say that φ is *axial*. We say that φ is *parabolic* if d_φ does not achieve a minimum in X.

Proposition 3.2. *An isometry φ of X is elliptic iff one of the following two equivalent conditions holds:*
(i) φ has a fixed point; *(ii) φ has a bounded orbit.*

Proof. It is obvious that φ is elliptic iff (i) holds and that (i) $\Rightarrow$ (ii). For the proof of (ii) $\Rightarrow$ (i), let $\overline{B}$ be the unique geodesic ball of smallest radius containing the bounded orbit $Y = \{\varphi^n(x) \mid n \in \mathbb{Z}\}$, see Proposition I.5.10. Now $\varphi(Y) = Y$, hence $\varphi(\overline{B}) = \overline{B}$ by the uniqueness of $\overline{B}$. Therefore φ fixes the center of $\overline{B}$. □

Proposition 3.3. *An isometry φ of X is axial iff there is a unit speed geodesic $\sigma : \mathbb{R} \to X$ and a number $t_0 > 0$ such that $\varphi(\sigma(t)) = \sigma(t + t_0)$ for all $t \in \mathbb{R}$. Such a geodesic σ will be called an axis of φ.*

Now let φ be axial, let $m_\varphi = \min d_\varphi > 0$ and set $A = A_\varphi = \{x \in X \mid d_\varphi(x) = m_\varphi\}$. Then A is closed, convex and isometric to $C \times \mathbb{R}$, where $C \subset A$ is closed and convex. Moreover, the axes of φ correspond precisely (except for the parameterization) to the geodesics $\{c\} \times \mathbb{R}$, $c \in C$.

Proof. Suppose φ is axial and let $x \in A$. Consider the geodesic segment ρ from x to $\varphi(x)$ and let y be the midpoint of ρ. Then $d(y, x) = d(y, \varphi(x)) = d(x, \varphi(x))/2$. Since $d(\varphi(y), \varphi(x)) = d(y, x)$, we get

$$d(y, \varphi(y)) \le d(y, \varphi(x)) + d(\varphi(x), \varphi(y)) \le d(x, \varphi(x)).$$

Now $x \in A$, hence $d(y, \varphi(y)) = d(x, \varphi(x))$. Hence the concatenation of ρ with $\varphi(\rho)$ is a geodesic segment. Therefore

$$\sigma = \cup_{n\in\mathbb{Z}} \varphi^n(\rho)$$

is an axis of φ. Hence there is an axis of φ through any $x \in A$. Clearly, any two axes of φ are parallel. Hence A is closed, convex and consists of a family of parallel geodesics. Therefore A is isometric to $C \times \mathbb{R}$ as claimed, where C is a closed and convex subset of X, see Proposition I.5.9. The other assertions are clear. □

Propostion 3.4. *If X is locally compact and if φ is a parabolic isometry of X, then there is a Busemann function $f = b(\xi, y, .)$ invariant under φ, that is, φ fixes $\xi \in X(\infty)$ and all horospheres centered at ξ are invariant under φ.*

Proof. Let $m = \inf d_\varphi \geq 0$. For $\delta > m$ set

$$X_\delta = \{x \in X \mid d_\varphi(x) \leq \delta\}.$$

Then X_δ is closed and convex since d_φ is convex. Moreover,

$$\cap_{\delta > m} X_\delta = \emptyset$$

since φ is parabolic. Let $x_0 \in X$ and set

$$f_\delta(x) = d(x, X_\delta) - d(x_0, X_\delta).$$

Then (i) $f_\delta(x_0) = 0$; (ii) f_δ is convex; (iii) f_δ has Lipschitz constant 1; (iv) for any $x \in X \backslash X_\delta$ and $r > 0$ with $r \leq d(x, X_\delta)$ there is a $z \in X$ with $d(x, z) = r$ and $f_\delta(x) - f_\delta(z) = r$. Now apply the Theorem of Arzela-Ascoli and Exercise 2.6. □

4. The cone at infinity and the Tits metric

Throughout this section we assume that X is a Hadamard space. For $\xi, \eta \in X(\infty)$ we define the *angle* by

$$\angle(\xi, \eta) = \sup_{x \in X} \angle_x(\xi, \eta).$$

Note that $\angle$ is a metric on $X(\infty)$ with values in $[0, \pi]$. The topology on $X(\infty)$ induced by this metric is, in general, different from the topology defined in the previous sections; to the latter we will refer as the *standard topology.*

4.1 Proposition. *$(X(\infty), \angle)$ is a complete metric space. Furthermore,*
(i) if $\xi_n \to \xi$ with respect to $\angle$, then $\xi_n \to \xi$ in the standard topology;
(ii) if $\xi_n \to \xi$ and $\eta_n \to \eta$ in the standard topology, then

$$\angle(\xi, \eta) \leq \liminf_{n \to \infty} \angle(\xi_n, \eta_n).$$

Proof. Let (ξ_n) be a Cauchy sequence in $X(\infty)$ with respect to $\angle$. Let $x \in X$ and let $\sigma_n = \sigma_{x,\xi_n}$ be the unit speed ray from x to ξ_n. Given $R > 0$ and $\varepsilon > 0$, there is $N = N(\varepsilon) \in \mathbb{N}$ such that $\angle(\xi_n, \xi_m) < \varepsilon$ for all $m, n \geq N$. Then

$$\angle_{\sigma_n(R)}(x, \xi_m) \geq \pi - \angle_{\sigma_n(R)}(\xi_n, \xi_m) \geq \pi - \angle(\xi_n, \xi_m) > \pi - \varepsilon.$$

By comparison with the Euclidean plane,

$$d_{x,R}(\xi_n, \xi_m) = d(\sigma_n(R), \sigma_m(R)) \leq 2R \tan \frac{\varepsilon}{2}.$$

Therefore (σ_n) converges to a unit speed ray σ and $\xi_n \to \xi := \sigma(\infty)$ with respect to the standard topology. Since $N = N(\varepsilon)$ does not depend on x and R, we also get $\xi_n \to \xi$ with respect to $\angle$. This proves (i) and that $\angle$ is complete. The proof of (ii) is similar. □

4.2 Proposition. *Let $\xi, \eta \in X(\infty)$. For $x \in X$, let $\sigma = \sigma_{x,\xi}$ be the ray from x to ξ. Then $\angle_{\sigma(t)}(\xi, \eta)$ and $\pi - \angle_{\sigma(t)}(x, \eta)$ are monotonically increasing with limit $\angle(\xi, \eta)$ as $t \to \infty$. If*

$$\angle(\xi, \eta) < \pi \quad \text{and} \quad \angle_x(\xi, \eta) = \angle(\xi, \eta),$$

then σ and the ray from x to η bound a flat convex region in X isometric to the convex hull of two rays in the flat plane with the same initial point and angle equal to $\angle(\xi, \eta)$.

Proof. Let $\gamma(t) = \angle_{\sigma(t)}(\xi, \eta)$ and $\varphi(t) = \angle_{\sigma(t)}(\eta, \sigma(0))$. Then we have $\gamma(t) + \varphi(t) \geq \pi$ and $\varphi(s) + \gamma(t) \leq \pi$ for $s > t$. Hence

$$\gamma(t) \leq \gamma(s) \quad \text{and} \quad \varphi(t) \geq \varphi(s)$$

for $s > t$. In particular, $\lim_{t\to\infty} \gamma(t)$ and $\lim_{t\to\infty} \varphi(t)$ exist. By definition we have $\lim_{t\to\infty} \gamma(t) \leq \angle(\xi, \eta)$. Now let $y \in X$ and let $\psi(t) = \angle_y(\sigma(t), \eta)$, $\delta(t) = \angle_{\sigma(t)}(y, \eta)$ and $\varepsilon(t) = \angle_{\sigma(t)}(y, \sigma(0))$. Then $\varepsilon(t) \to 0$ and $\psi(t) \to \angle_y(\xi, \eta)$ for $t \to \infty$. Now

$$\psi(t) + \delta(t) \leq \pi \quad \text{and} \quad \varepsilon(t) + \delta(t) + \gamma(t) \geq \pi,$$

hence $\psi(t) \leq \varepsilon(t) + \gamma(t)$, hence $\angle_y(\xi, \eta) \leq \lim_{t\to\infty} \gamma(t)$.

This proves the assertion about $\gamma(t)$. Now

$$\pi - \varphi(t) \leq \gamma(t) \leq \pi - \varphi(s)$$

for $s > t$, hence we also obtain the assertion $\lim_{t\to\infty}(\pi - \varphi(t)) = \angle(\xi, \eta)$. The last assertion of the lemma follows from the equality discussion in assertion (i) of Corollary I.5.8 since $\gamma(t) \equiv$ constant in that case, and hence $\varphi(s) = \pi - \gamma(t)$ for $s > t$. □

4.3 Exercise. Let $\xi, \eta \in X(\infty)$ and assume $\xi \neq \eta$. Let σ_0 and σ_1 be the unit speed rays from x to ξ and η respectively. Define angles

$$\alpha_{s,t} = \angle_{\sigma_1(s)}(\sigma_0(t), x), \quad \beta_{s,t} = \angle_{\sigma_0(t)}(\sigma_1(s), x).$$

Prove: $\pi - \alpha_{s,t} - \beta_{s,t}$ is monotonically increasing as s, t increase and

$$\angle(\xi, \eta) = \lim_{s,t\to\infty} \pi - \alpha_{s,t} - \beta_{s,t}.$$

The second assertion also follows from the monotonicity and Theorem 4.4 below.

4.4 Theorem. *Let $\xi \neq \eta \in X(\infty)$ and $a, b > 0$. For $x \in X$ let σ_0 and σ_1 be the unit speed rays from x to ξ and η respectively. Then $c = \lim_{t\to\infty} d(\sigma_0(at), \sigma_1(bt))/t$ is independent of the choice of x. In the Euclidean triangle with sides a, b, c and angles α, β, γ in the opposite vertices A, B, C, see Figure 1, we have $\gamma = \angle(\xi, \eta)$ and*

$$\alpha_t = \angle_{\sigma_1(bt)}(\sigma_0(at), x) \to \alpha, \quad \beta_t = \angle_{\sigma_0(at)}(\sigma_1(bt), x) \to \beta.$$

For $s \in (0,1)$ given and E the point on c with $d(E,B) = sc$ we have

$$\lim_{t\to\infty} \frac{1}{t} d(x, \rho_t(s)) = e := d(C,E),$$

where $\rho_t : [0,1] \to X$ is the geodesic from $\sigma_0(at)$ to $\sigma_1(bt)$. Furthermore, if $e \neq 0$, then the geodesics $\sigma_{s,t} : [0, et] \to X$ with $\sigma_{s,t}(0) = x$ and $\sigma_{s,t}(et) = \rho_s(t)$ converge, for $t \to \infty$, to a unit speed ray σ_s from x such that $\rho_t(s) \to \sigma_s(\infty)$, and $\zeta = \sigma_s(\infty)$ is the unique point in $X(\infty)$ with

$$\angle(\zeta,\xi) = \angle_C(E,B), \quad \angle(\zeta,\eta) = \angle_C(E,A).$$

Moreover,

$$\lim_{t\to\infty} \angle_{\rho_t(s)}(\sigma_0(at), x) = \alpha' := \angle_E(B,C),$$
$$\lim_{t\to\infty} \angle_{\rho_t(s)}(\sigma_1(bt), x) = \beta' := \angle_E(A,C).$$

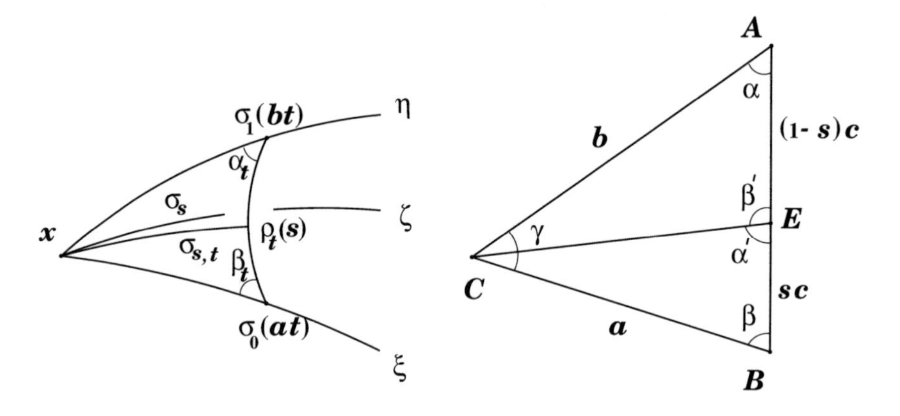

FIGURE 1

Proof. The independence of c from x is immediate from the definition of asymptoticity. For $\gamma_x = \angle_x(\xi,\eta)$ and $c(t) = L(\sigma_t) = d(\sigma_0(at), \sigma_1(bt))$, we have

$$\begin{aligned}
&(a^2+b^2-2ab\cos\gamma_x)^{\frac{1}{2}} \leq \frac{c(t)}{t} \leq a\cos\beta_t + b\cos\alpha_t \\
&= c_{t,E}\cos(\alpha_t - \alpha_{t,E}) = c_{t,E}\cos(\beta_t - \beta_{t,E}) \leq c_{t,E} \\
&= (a^2+b^2-2ab\cos(\pi-\alpha_t-\beta_t))^{\frac{1}{2}} \leq (a^2+b^2-2ab\cos\angle(\xi,\eta))^{\frac{1}{2}},
\end{aligned}$$

where the inequalities in the first line follow from the Cosine Inequalities, see Proposition I.5.2, the equalities in the second line from Proposition I.5.3, where

$\alpha_{t,E}$ and $\beta_{t,E}$ are the angles and $c_{t,E}$ is the length of the third edge in the Euclidean triangle with two edges of length a and b subtending an angle $\pi - \alpha_t - \beta_t$, and the inequality in the third line since

$$\begin{aligned}\pi - \alpha_t - \beta_t &\leq \angle_{\sigma_1(bt)}(\sigma_0(at), \eta) + \angle_{\sigma_0(at)}(\sigma_1(bt), \xi) - \pi \\ &\leq \angle_{\sigma_0(at)}(\xi, \eta) \leq \angle(\xi, \eta).\end{aligned}$$

Now $c(t)/t$ tends to c and is independent of x. Since $\angle(\xi, \eta) = \sup_{x \in X} \gamma_x$ we get

$$c^2 = a^2 + b^2 - 2ab \cos \angle(\xi, \eta)$$

and therefore $\angle(\xi, \eta) = \gamma$. Furthermore, we have $\pi - \alpha_t - \beta_t \to \angle(\xi, \eta)$ and $c_{t,E} \to c$ as $t \to \infty$. Now $c \neq 0$ and hence $\alpha_{t,E} \to \alpha$, $\beta_{t,E} \to \beta$. We conclude that $\alpha_t \to \alpha$, $\beta_t \to \beta$ as $t \to \infty$. We also have

$$d(x, \rho_t(s)) \leq d(\overline{x}, \overline{\rho}_t(s)),$$

where $\overline{\rho}_t : [0,1] \to \mathbb{R}^2$ is a line of length $c(t)$ and $d(\overline{x}, \overline{\rho}_t(0)) = at$, $d(\overline{x}, \overline{\rho}_t(1)) = bt$. Now $d(\overline{x}, \overline{\rho}_t(s))/t \to e$ as $t \to \infty$, therefore

$$\limsup_{t \to \infty} \frac{1}{t} d(x, \rho_t(s)) \leq e.$$

On the other hand, we have

$$\frac{d^2(x, \rho_t(s))}{t^2} \geq a^2 + s^2 \frac{c(t)^2}{t^2} - 2as\frac{c(t)}{t} \cos(\alpha_t)$$

by the First Cosine Inequality. Since $\alpha_t \to \alpha$ and $c(t)/t \to c$ we conclude that

$$\liminf_{t \to \infty} \frac{1}{t} d(x, \rho_t(s)) \geq e$$

and hence that $d(x, \rho_t(s))/t \to e$ at $t \to \infty$.

We now prove that the geodesics $\sigma_{s,t}$ converge to a ray as claimed. Given $\varepsilon > 0$, there is $T \geq 0$ such that

$$|\frac{c(t)^2}{t^2} - c^2|, \ |2as\frac{c(t)}{t} \cos \alpha_t - 2asc \cos \alpha| < \frac{\varepsilon}{2}$$

for all $t \geq T$. Then we obtain for $t > t' \geq T$ and $r = t'/t$,

$$\begin{aligned}\frac{d^2(\sigma_0(at'), \rho_t(s))}{t^2} &\geq \frac{1}{t^2}(a^2(t-t')^2 + s^2 c(t)^2 - 2a(t-t')sc(t) \cos \alpha_t) \\ &\geq a^2(1-r)^2 + s^2c^2 - 2a(1-r)sc \cos \alpha - \varepsilon.\end{aligned}$$

Now denote by $\overline{\alpha}_t$ the angle in the Euclidean triangle with side lengths $a, b, c(t)/t$ corresponding to α_t. Then $\overline{\alpha}_t \geq \alpha_t$ and $\overline{\alpha}_t \to \alpha$. By enlarging T if necessary, we also have

$$|2as\frac{c(t)}{t}\cos\overline{\alpha}_t - 2asc\cos\alpha| < \frac{\varepsilon}{2}$$

for all $t \geq T$. Since triangles in X are CAT_0 we obtain, for $t > t' \geq T$,

$$\begin{aligned}\frac{d^2(\sigma_0(at'), \rho_t(s))}{t^2} &\leq \frac{1}{t^2}(a^2(t-t')^2 + s^2c(t)^2 - 2a(t-t')sc(t)\cos\overline{\alpha}_t) \\ &\leq a^2(1-r)^2 + s^2c^2 - 2a(1-r)sc\cos\alpha + \varepsilon.\end{aligned}$$

An analogous estimate holds for $d(\sigma_1(bt'), \rho_t(s))$. Euclidean comparison for the triangle $(\sigma_0(at'), \sigma_1(bt'), \rho_t(s))$ shows that

$$d(\rho_t'(s), \rho_t(s)) \leq ((1-r)e + \varepsilon)t$$

for all t, t' sufficiently large, where $r = t'/t$. On the other hand, $d(x, \rho_t(s))/t \to e$ and hence the triangle inequality implies

$$d(\rho_t'(s), \rho_t(s)) \geq ((1-r)e - \varepsilon)t$$

for all t, t' sufficiently large. In particular, up to the term εt (and ε can be chosen arbitrarily small), the concatenation of the geodesic from x to $\rho_{t'}(s)$ with the geodesic from $\rho_{t'}(s)$ to $\rho_t(s)$ is a shortest connection from x to $\rho_t(s)$. By comparison with Euclidean geometry we conclude that for any $R \geq 0$, the distance $d(\sigma_{s,t}(R), \sigma_{s,t'}(R))$ is arbitrarily small for all t, t' sufficiently large. Hence the geodesic segments $\sigma_{s,t}$ converge to a ray σ_s. Note that σ_s has unit speed since $d(x, \rho_t(s))/t \to e$ and that $\rho_t(s) \to \sigma_s(\infty) = \zeta$ by the very definition of the topology of $\overline{X}$.

For t large, the triangle with vertices $x, \sigma_0(at), \rho_t(s)$ has sides of length approximately equal to at, sct, et. By comparison we get

$$\angle_x(\xi, \zeta) = \lim_{t\to\infty} \angle_x(\sigma_0(at), \rho_t(s)) \leq \angle_C(B, E).$$

Similarly $\angle_x(\eta, \zeta) \leq \angle_C(A, E)$. Clearly, ζ is independent of the choice of x and hence

$$(*) \qquad \angle(\xi, \zeta) \leq \angle_C(B, E), \quad \angle(\eta, \zeta) \leq \angle_C(A, E).$$

Now

$$\angle(\xi, \zeta) + \angle(\eta, \zeta) \geq \angle(\xi, \eta) = \angle_C(A, B),$$

hence we have equality in $(*)$.

Suppose now that $\zeta' \in X(\infty)$ is a point with

$$(**) \qquad \angle(\xi, \zeta') = \angle_C(B, E), \quad \angle(\eta, \zeta') = \angle_C(A, E),$$

and let σ' be the unit speed ray from x to ζ'. If $\zeta \neq \zeta'$, then there is an $\varepsilon > 0$ such that $d(\sigma'(et), \rho_t(s)) \geq \varepsilon t$ for all t sufficiently large. By comparison with Euclidean geometry we get that at least one of the following inequalities holds:

$$c_0^2 = \lim_{t\to\infty} \frac{1}{t^2} d^2(\sigma_0(at), \sigma'(et)) \geq s^2c^2 + \varepsilon^2 \text{ or}$$
$$c_1^2 = \lim_{t\to\infty} \frac{1}{t^2} d^2(\sigma_1(bt), \sigma'(et)) \geq (1-s)^2c^2 + \varepsilon^2.$$

However, by the first part of the proof we have

$$\angle(\xi, \zeta') = \frac{a^2 + e^2 - c_0^2}{2ae}, \quad \angle(\eta, \zeta') = \frac{b^2 + e^2 - c_1^2}{2be}.$$

Hence (**) implies $c_0^2 = s^2c^2$ and $c_1^2 = (1-s)^2c^2$. We conclude $\zeta = \zeta'$.

As for the last claim, let

$$\alpha(t) = \angle_{\rho_t(s)}(\sigma_0(at), x), \quad \beta(t) = \angle_{\rho_t(s)}(x, \sigma_1(bt)).$$

Then $\alpha(t) + \beta(t) \geq \pi$. By comparison with Euclidean geometry, applied to the triangles $\Delta(\rho_t(s), x, \sigma_0(at))$ and $\Delta(\rho_t(s), x, \sigma_1(bt))$ we get

$$\limsup_{t\to\infty} \alpha(t) \leq \alpha' \quad \text{and} \quad \limsup_{t\to\infty} \beta(t) \leq \beta'.$$

Now $\alpha' + \beta' = \pi$, hence the claim. □

4.5 Exercise. Use the notation of Theorem 4.4 and assume that $e \neq 0$. Let F be a point on the edge of length a on the Euclidean triangle such that $d(C, F) = ra$ for some $r \in (0, 1)$. Prove:

$$d(E, F) = \lim_{t\to\infty} \frac{1}{t} d(\rho_t(s), \sigma_0(rat))$$

and corresponding assertions about the angles at F.

4.6 Corollary. *If $\xi, \eta \in X(\infty)$ are points with $\angle(\xi, \eta) < \pi$, then there is a unique minimizing geodesic in $X(\infty)$ (with respect to $\angle$) from ξ to η.*

Proof. The existence of a geodesic from ζ to η follows from Proposition I.1.5 and Theorem 4.4, where we put $a = b = 0$, $s = 1/2$. The uniqueness follows from the uniqueness of the point ζ in Theorem 4.4. □

The *cone at infinity*, $C_\infty X$, is the set $[0, \infty) \times X(\infty)/\sim$, where $\{0\} \times X(\infty)$ shrinks to one point, together with the metric d_∞ defined by

$$d_\infty((a, \xi), (b, \eta)) = a^2 + b^2 - 2ab\cos\angle(\xi, \eta). \tag{4.7}$$

That is, $(C_\infty X, d_\infty)$ is the *Euclidean cone* over $X(\infty)$. We may also think of $C_\infty X$ as the set of equivalence classes of geodesic rays, where we do not require unit speed and where we say that geodesic rays σ_0 and σ_1 are equivalent if $d(\sigma_0(t), \sigma_1(t))$ is uniformly bounded in t. In the notation of Theorem 4.4 we have

$$d_\infty((a,\xi),(b,\eta)) = \lim_{t\to\infty} \frac{1}{t} d(\sigma_0(at), \sigma_1(bt)).$$

In this sense, the geometry of X with respect to the rescaled metric d/t converges to the geometry of $C_\infty X$ as $t \to \infty$.

Given $x \in X$ and a sequence of geodesics $\sigma_n : [0, t_n] \to X$ with speed $v_n \geq 0$ and $\sigma_n(0) = x$, we say that $\sigma_n \to (a, \xi)$ if $t_n \to \infty$, if $v_n \to a$ and, if $a \neq 0$, if $\sigma_n(t_n) \to \xi$.

Theorem 4.8. *$(C_\infty X, d_\infty)$ is a Hadamard space.*

Proof. Note that $C_\infty X$ is simply connected since we can contract $C_\infty X$ onto the distinguished point $x_\infty = [\{0\} \times X(\infty)]$. The metric d_∞ is complete since $\angle$ is a complete metric on $X(\infty)$.

We show now that d_∞ is a geodesic metric. To that end, let $(a,\xi), (b,\eta) \in C_\infty X$. We want to show that there is a minimizing geodesic between these two points. The only non-trivial case is $a, b > 0$ and $0 < \angle(\xi,\eta) < \pi$. Choose $x \in X$ and let σ_0, σ_1 be the unit speed rays from x with $\sigma_0(\infty) = \xi$ and $\sigma_1(\infty) = \eta$. Then, in the notation of Theorem 4.4, the geodesics $\sigma_{s,t}$ converge to geodesic rays σ_s of speed $e_s > 0$, $0 \leq s \leq 1$, and, by Theorem 4.4 and the definition of d_∞, the curve $\rho(s) = (e_s, \sigma_s(\infty)), 0 \leq s \leq 1$, is a minimizing geodesic form (a,ξ) to (b,η). It is immediate from Corollary 4.6 (or Theorem 4.4) that ρ is the unique minimizing geodesic from (a,ξ) to (b,η).

It remains to show that $C_\infty X$ is nonpositively curved. By Lemma I.3.4, it suffices to consider triangles in $C_\infty X$ with minimizing sides. Let Δ be such a triangle and let (a_i, ξ_i), $1 \leq i \leq 3$, be the vertices of Δ. The only non-trivial case is $a_i > 0$ and $0 < \angle(\xi_i, \xi_j) < \pi$ for $i \neq j$. Recall that the minimal geodesics $\rho_i : [0,1] \to C_\infty X$ from (a_i, ξ_i) to (a_{i+1}, ξ_{i+1}) are unique (indices are counted mod 3).

Let $(e_1, \zeta_1) = \rho_i(s_1)$ and $(e_2, \zeta_2) = \rho_j(s_2)$, where $i \neq j$ and $0 < s_1, s_2 < 1$. Let σ_1 and σ_2 be the unit speed rays from a given point $x \in X$ to ζ_1 and ζ_2 respectively. By Theorem 4.4,

$$d_\infty((e_1,\zeta_1),(e_2,\zeta_2)) = \lim_{R\to\infty} \frac{1}{R} d(\sigma_1(e_1 R), \sigma_2(e_2 R)).$$

Let $\tilde\sigma_k$ be the unit speed ray from x to ξ_k, $1 \leq k \leq 3$. If $\rho_{k,t} : [0,1] \to X$ denotes the geodesic from $\tilde\sigma_k(a_k t)$ to $\tilde\sigma_{k+1}(a_{k+1} t)$, and if $\hat\sigma_{1,t}$ respectively $\hat\sigma_{2,t}$ are the geodesics from x with $\hat\sigma_{1,t}(e_1 t) = \rho_{i,t}(s_1)$ respectively $\hat\sigma_{2,t}(e_2 t) = \rho_{j,t}(s_2)$, then, by Theorem 4.4,

$$\sigma_k(e_k R) = \lim_{t\to\infty} \hat\sigma_{k,t}(e_k R),\ k = 1, 2.$$

For t large, the triangle $(\rho_{1,t}, \rho_{2,t}, \rho_{3,t})$ has sides of length approximately tl_k, where $l_k = L(\rho_k)$. Now let $(\overline{\rho}_1, \overline{\rho}_2, \overline{\rho}_3)$ be the Euclidean comparison triangle of (ρ_1, ρ_2, ρ_3) and let $d_E = d(\overline{\rho}_i(s_1), \overline{\rho}_j(s_2))$. Since X is a Hadamard space we conclude

$$d(\rho_{i,t}(s_1), \rho_{j,t}(s_2)) \leq (d_E + \varepsilon)t$$

where $\varepsilon > 0$ is given and t is sufficiently large. Hence

$$\begin{aligned}\frac{1}{R} d(\sigma_1(e_1 R), \sigma_2(e_2 R)) &= \lim_{t\to\infty} \frac{1}{R} d(\hat{\sigma}_{1,t}(e_1 R), \hat{\sigma}_{2,t}(e_2 R)) \\ &\leq \limsup_{t\to\infty} \frac{1}{t} d(\rho_{i,t}(s_1), \rho_{j,t}(s_2)) \leq d_E.\end{aligned}$$

It follows that triangles in $C_\infty X$ are CAT_0 and hence that $C_\infty X$ is a Hadamard space. □

4.9 Corollary. *$(X(\infty), \angle)$ has curvature ≤ 1. More precisely,*
(i) geodesics in $X(\infty)$ of length $< \pi$ are minimal;
(ii) triangles in $X(\infty)$ of perimeter $< 2\pi$ are CAT_1.

Proof. The relation between distances in $C_\infty X$ and angles in $X(\infty)$ is exactly as the relation between distances in Euclidean space E^3 and angles in the unit sphere S^2. □

4.10 Lemma. *Let X be a locally compact Hadamard space. In the notation of Theorem 4.4 let $a = b = 1$ and assume that there is no geodesic in X from ξ to η. Let $\rho_t(s_t)$ be the point on ρ_t closest to x. Then $d(x, \rho_t(s_t)) \to \infty$. If $\zeta \in X(\infty)$ is an accumulation point of $\rho_t(s_t)$ for $t \to \infty$, then*

$$\angle(\xi, \zeta) = \angle(\zeta, \eta) = \gamma/2.$$

Proof. Since X is locally compact and since there is no geodesic from ξ to η we have $d(x, \rho_t) \to \infty$ for $t \to \infty$. From Theorem 4.4 we conclude $d(x, \rho_t(1/2))/t \to \cos(\gamma/2) < 1$ and hence $s_t \neq 0, 1$. Therefore

$$\angle_{\rho_t(s_t)}(x, \sigma_0(t)),\ \angle_{\rho_t(s_t)}(x, \sigma_1(t)) \geq \pi/2.$$

Since (in the notation of Theorem 4.4) $\alpha_t, \beta_t \to (\pi - \gamma)/2$ we get

$$\angle_x(\xi, \zeta),\ \angle_x(\zeta, \eta) \leq \gamma/2.$$

The triangle inequality implies that we must have equality in these inequalities. □

Following Gromov, we denote by Td, the *Tits metric*, the interior metric on $X(\infty)$ associated to the angle metric $\angle$. The name derives from its close relationship with the Tits building associated to symmetric spaces. By definition, we have $Td(\xi, \eta) \geq \angle(\xi, \eta)$ for all $\xi, \eta \in X(\infty)$.

4.11 Theorem. *Assume that X is a locally compact Hadamard space. Then each connected component of $(X(\infty), Td)$ is a complete geodesic space with curvature ≤ 1. Furthermore, for $\xi, \eta \in X(\infty)$ the following hold:*

(i) if there is no geodesic in X from ξ to η, then $Td(\xi,\eta) = \angle(\xi,\eta) \leq \pi$;

(ii) if $\angle(\xi,\eta) < \pi$, then there is no geodesic in X from ξ to η and there is a unique minimizing Td-geodesic in $X(\infty)$ from ξ to η; Td-triangles in X of perimeter $< 2\pi$ are CAT_1;

(iii) if there is a geodesic σ in X from ξ to η, then $Td(\xi,\eta) \geq \pi$, and equality holds iff σ bounds a flat half plane;

(iv) if $(\xi_n), (\eta_n)$ are sequences in $X(\infty)$ such that $\xi_n \to \xi$ and $\eta_n \to \eta$ in the standard topology, then $Td(\xi,\eta) \leq \liminf_{n\to\infty} Td(\xi_n,\eta_n)$.

Proof. The first assertion in (ii) and (iii) respectively is clear since $\angle(\xi,\eta) = \pi$ if there is a geodesic in X from ξ to η. If there is no geodesic in X from ξ to η, then, by Lemma 4.10, there is a midpoint ζ in $X(\infty)$ between them with respect to the $\angle$-metric. Now the $\angle$-metric on $X(\infty)$ is complete, see Proposition 4.1. Hence we can construct an $\angle$-geodesic from ξ to η by choosing midpoints iteratively. Since $\angle$-geodesics are Td-geodesics, we obtain (i). The uniqueness assertion in (ii) follows from the corresponding uniqueness assertion of midpoints in Theorem 4.4. The completeness of Td also follows since the $\angle$-metric is complete, see Proposition 4.1. We now prove (iv) and that each connected component of $(X(\infty), Td)$ is geodesic. By the corresponding semicontinuity of $\angle$, see Proposition 4.1, we conclude from (i) that (iv) holds if $\liminf_{n\to\infty} Td(\xi_n,\eta_n) < \pi$. From (ii) we also get that for any pair of points $\xi, \eta \in X(\infty)$ with $Td(\xi,\eta) < \pi$ there is a minimizing Td-geodesic from ξ to η. By induction we assume that this also holds when we replace π by $k\pi$, $k \geq 1$. Now suppose $\liminf_{n\to\infty} Td(\xi_n,\eta_n) < (k+1)\pi$. By th definition of Td, there is a point ζ_n almost half way between ξ_n and η_n. Passing to a subsequence if necessary, we assume $\zeta_n \to \zeta$ in the standard topology and obtain, by induction,

$$
\begin{aligned}
Td(\xi,\eta) &\leq Td(\xi,\zeta) + Td(\zeta,\eta) \\
&\leq \liminf_{n\to\infty} Td(\xi_n,\zeta_n) + \liminf_{n\to\infty} Td(\zeta_n,\eta_n) \leq \liminf_{n\to\infty} Td(\xi_n,\eta_n).
\end{aligned}
$$

In particular, $Td(\xi,\eta) < (k+1)\pi$. There exists (another) sequence (ζ_n) such that

$$Td(\xi,\zeta_n), Td(\eta,\zeta_n) \to Td(\xi,\eta)/2,$$

hence, by the semicontinuity, there is a point ζ with

$$Td(\xi,\zeta) = Td(\eta,\zeta) = Td(\xi,\eta)/2.$$

By the inductive hypothesis, there are minimizing Td-geodesics from ξ to ζ and ζ to η. Their concatenation is a minimizing Td-geodesic from ξ to η. Hence we conclude (iv) and that the components of $(X(\infty), Td)$ are geodesic spaces.

As for the statement about flat half planes in (iii), let ζ be a point with $Td(\zeta, \xi) = Td(\zeta, \eta) = \pi/2$. Since

$$\angle_x(\zeta, \xi) + \angle_x(\zeta, \eta) \geq \pi$$

for any x on the geodesic (in X) from ξ to η and $\angle_x \leq Td$, we have

$$\angle_x(\zeta, \xi) = \angle_x(\zeta, \eta) = \pi/2$$

for all such x. Now the rigidity part of Lemma I.5.8(i) applies.

The claims about the curvature bound and about triangles in (ii) are immediate from Corollary 4.9. $\square$

CHAPTER III

Weak Hyperbolicity

One of our main objectives in this chapter is the discussion of transitivity properties of the geodesic flow and related topics. The first two sections are greatly influenced by the work of Eberlein on geodesic flows in [Eb2, Eb3]. The third section deals with the main results of the author's thesis (published in [Ba1]), but here in the context of Hadamard spaces instead of Hadamard manifolds. The fourth section deals with applications to harmonic functions and random walks on countable groups, compare [Ba3, BaL1]. Many of the ideas in Section 4 go back to Furstenberg [Fu1, Fu2, Fu3].

Let X be a Hadamard space. We say that X is *geodesically complete* if every geodesic segment of X is part of a *complete geodesic* of X, that is, a geodesic which is defined on the whole real line. For X geodesically complete, we denote by GX the set of unit speed geodesics of X, equipped with the topology of uniform convergence on bounded subsets. The *geodesic flow* (g^t) acts on GX by reparameterization,

$$g^t(\sigma)(s) = \sigma(s+t) .$$

If X is a smooth manifold, then

$$GX \to SX; \ \sigma \to \dot{\sigma}(0)$$

defines a homeomorphism from GX to the unit tangent bundle SX of X. With respect to this homeomorphism, the geodesic flow above corresponds to the usual geodesic flow on SX (also denoted g^t),

$$g^t(v) = \dot{\gamma}_v(t), \ v \in SX,$$

where γ_v denotes the geodesic in X with $\dot{\gamma}_v(0) = v$.

1. The duality condition

In our considerations it will be important to have a large group of isometries. Let X be a Hadamard space and Γ a group of isometries of X. Following Eberlein [Eb2], we say that $\xi, \eta \in X(\infty)$ are Γ-*dual* if there is a sequence (φ_n) in Γ such that $\varphi_n(x) \to \xi$ and $\varphi_n^{-1}(x) \to \eta$ as $n \to \infty$ for some (and hence any) $x \in X$.

Duality is a symmetric relation. The set of points Γ-dual to a given point in $X(\infty)$ is closed and Γ-invariant. We say that Γ satisfies the *duality condition* if for any unit speed geodesic $\sigma : \mathbb{R} \to X$ the endpoints $\sigma(-\infty)$ and $\sigma(\infty)$ are Γ-dual. The duality condition was introduced by Chen and Eberlein [CE1]. The condition is somewhat mysterious, at least as far as Hadamard spaces are concerned (as opposed to Hadamard manifolds). May be it is only reasonable for geodesically complete Hadamard spaces.

1.1 Lemma [Eb2]. *Let X be a geodesically complete Hadamard space, and let $\sigma, \rho : \mathbb{R} \to X$ be unit speed geodesics in X. Let (φ_n) be a sequence of isometries of X such that $\varphi_n^{-1}(x) \to \sigma(\infty)$ and $\varphi_n(x) \to \rho(-\infty)$ as $n \to \infty$ for some (and hence any) $x \in X$. Then there are sequences (σ_n) in GX and (t_n) in $\mathbb{R}$ such that $\sigma_n \to \sigma$, $t_n \to \infty$ and $\varphi_n \circ g^{t_n}(\sigma_n) \to \rho$ as $n \to \infty$.*

1.2 Remarks. (a) The converse to Lemma 1.1 is clear: let $\sigma, \rho \in GX$ and suppose there are sequences (σ_n) in GX, (t_n) in $\mathbb{R}$ and (φ_n) of isometries of X such that $\sigma_n \to \sigma$, $t_n \to \infty$ and $\varphi_n \circ g^{t_n}(\sigma_n) \to \sigma$ as $n \to \infty$. Then $\varphi_n^{-1}(x) \to \sigma(\infty)$ and $\varphi_n(x) \to \rho(-\infty)$ as $n \to \infty$ for one (and hence any) $x \in X$.

(b) Note that the assumption on the sequence (φ_n) in Lemma 1.1 only specifies $\sigma(\infty)$ and $\rho(-\infty)$. Thus we may replace σ by any asymptotic $\sigma' \in GX$ and ρ by any negatively asymptotic $\rho' \in GX$, keeping the sequence (φ_n) fixed.

Since $\sigma_n \to \sigma$ and $\varphi_n^{-1} \circ g^{t_n}(\sigma_n) \to \rho$ we have

$$t_n - d(\sigma(0), \varphi_n^{-1}(\rho(0))) \to 0 \quad \text{as} \quad n \to \infty .$$

For σ' asymptotic to σ we also have

$$d(\sigma'(0), \varphi_n^{-1}(\rho(0))) - d(\sigma(0), \varphi_n^{-1}(\rho(0))) \;\to\; b(\sigma(\infty), \sigma(0), \sigma'(0))$$

as $n \to \infty$. Hence we may replace the sequence (t_n) by the sequence (t'_n) given by

$$t'_n \;=\; t_n + b(\sigma(\infty), \sigma(0), \sigma'(0)), \; n \in \mathbb{N},$$

when replacing σ by σ'. A similar remark applies to geodesics $\rho' \in GX$ negatively asymptotic to ρ.

Proof of Lemma 1.1. For $m \geq 0$ and $n \in \mathbb{N}$ let $\sigma_{n,m} : \mathbb{R} \to X$ be a unit speed geodesic with

$$\sigma_{n,m}(-m) \;=\; \sigma(-m) \quad \text{and} \quad \sigma_{n,m}(t_{n,m} + m) \;=\; \varphi_n^{-1}(\rho(m)),$$

where

$$t_{n,m} \;=\; d(\sigma(-m), \varphi_n^{-1}(\rho(m))) - 2m .$$

Set $N(0) = 1$ and define $N(m)$, $m \geq 2$, recursively to be the smallest number $> N(m-1)$ such that

$$d(\sigma_{n,m}(m), \sigma(m)) \;\leq\; \frac{1}{m} \quad \text{and} \quad d(\varphi_n \circ \sigma_{n,m}(t_{n,m} - m), \rho(-m)) \;\leq\; \frac{1}{m}$$

for all $n \geq N(m)$. Note that such a number $N(m)$ exists since $\varphi_n^{-1}(\rho(m)) \to \sigma(\infty)$ and $\varphi_n(\sigma(-m)) \to \rho(-\infty)$. Now set $\sigma_n := \sigma_{n,N(m)}$ for $N(m) \leq n < N(m+1)$. $\square$

1.3 Corollary. *Let X be a geodesically complete Hadamard space and Γ a group of isometries of X satisfying the duality condition. Let $\sigma, \rho : \mathbb{R} \to X$ be unit speed geodesics in X with $\sigma(\infty) = \rho(\infty)$. Then there are sequences (σ_n) in GX, (t_n) in $\mathbb{R}$ and (φ_n) in Γ such that $\sigma_n \to \sigma$, $t_n \to \infty$ and $\varphi_n \circ g^{t_n}(\sigma_n) \to \rho$ as $n \to \infty$.*

Proof. Since Γ satisfies the duality condition, there is a sequence (ψ_n) of isometries in Γ such that $\psi_n(x) \to \rho(\infty)$ and $\psi_n^{-1}(x) \to \rho(-\infty)$ as $n \to \infty$. Now $\sigma(\infty) = \rho(\infty)$, and hence Lemma 1.1 applies. □

We say that a geodesic $\sigma \in GX$ is *nonwandering* mod Γ (with respect to the geodesic flow) if there are sequences (σ_n) in GX, (t_n) in $\mathbb{R}$ and (φ_n) in Γ such that $\sigma_n \to \sigma$, $t_n \to \infty$ and $\varphi_n \circ g^{t_n}(\sigma_n) \to \sigma$. The set of geodesics in GX that are nonwandering mod Γ is closed in GX and invariant under the geodesic flow and Γ. It is clear that the endpoints of a geodesic $\sigma \in GX$ are Γ-dual if σ is nonwandering mod Γ. Vice versa, Lemma 1.1, applied in the case where X is geodesically complete and $\sigma = \rho$, shows that a geodesic $\sigma \in GX$ is nonwandering mod Γ if the endpoints of σ are Γ-dual.

1.4 Corollary. *Let X be a geodesically complete Hadamard space and Γ a group of isometries of X. Then Γ satisfies the duality condition if and only if every $\sigma \in GX$ is nonwandering mod Γ.* □

We say that a geodesic $\sigma \in GX$ is Γ-*recurrent* if there exist sequences (t_n) in $\mathbb{R}$ and (φ_n) in Γ such that $t_n \to \infty$ and $\varphi_n \circ g^{t_n}(\sigma) \to \sigma$ as $n \to \infty$. The set of Γ-recurrent geodesics in GX is invariant under the geodesic flow and Γ.

1.5 Corollary. *Let X be a geodesically complete, separable Hadamard space and Γ a group of isometries of X. Then Γ satisfies the duality condition if and only if a dense set of geodesics $\sigma \in GX$ is Γ-recurrent.*

Proof. Since X is separable, the topology of X,and hence also of GX, has a countable base. Now Corollary 1.4 implies that the set of Γ-recurrent geodesics is a dense G_δ-subset of GX if Γ satisfies the duality condition. The other direction is clear. □

If X is a Hadamard manifold, then the Liouville measure on SX is invariant under the geodesic flow. Hence, if Γ is a group of isometries of X that acts properly discontinuously and such that $\mathrm{vol}(X/\Gamma) < \infty$, then Γ satisfies the duality condition by the Poincaré recurrence theorem and Corollary 1.4. However, for a general Hadamard space X, there is (so far) no natural invariant measure on GX with good properties.

1.6 Question. Let X be a geodesically complete Hadamard space and Γ a group of isometries of X. Does Γ satisfy the duality condition if it acts properly discontinuously and cocompactly?

1.7 Problem/Exercise. Let X be a geodesically complete Hadamard space. Construct measures on GX which are invariant under isometries and the geodesic flow of X.

In the case of 2-dimensional polyhedra with piecewise smooth metrics, the answer to Question 1.6 is affirmative, see [BB3]. In this reference, one also finds a natural generalization of the Liouville measure for geodesic flows on polyhedra with piecewise smooth metrics.

The following lemma and proposition are taken from [CE1] and [BBE] respectively.

1.8 Lemma. *Suppose $\xi, \eta \in X(\infty)$ are the endpoints of a geodesic $\sigma : \mathbb{R} \to X$. Let (φ_n) be a sequence of isometries of X such that $\varphi_n(x) \to \xi$ and $\varphi_n^{-1}(x) \to \zeta \in X(\infty)$ as $n \to \infty$. Then $\varphi_n^{-1}(\eta) \to \zeta$ as $n \to \infty$.*

Proof. Let $x = \sigma(0)$. Then $\angle_x(\xi, \varphi_n(x)) \to 0$ and hence

$$\angle_x(\varphi_n(x), \eta) \;\geq\; \pi - \angle_x(\xi, \varphi_n(x)) \;\to\; \pi\,.$$

By comparison with Euclidean geometry we get, for the metric $d_{x,R}$ as in (II.2.3),

$$d_{x,R}(\varphi_n^{-1}(x), \varphi_n^{-1}(\eta)) \;=\; d_{\varphi_n(x),R}(x, \eta) \;\to\; 0\,,$$

where $R > 0$ is arbitrary. Since $\varphi_n^{-1}(x) \to \zeta$ we conclude $\varphi_n^{-1}(\eta) \to \zeta$. $\square$

1.9 Proposition. *Suppose X is geodesically complete and Γ satisfies the duality condition. Let $\xi \in X(\infty)$. Then Γ operates minimally on the closure $\overline{\Gamma(\xi)}$ of the Γ-orbit of ξ, that is, $\overline{\Gamma(\zeta)} = \overline{\Gamma(\xi)}$ for any $\zeta \in \overline{\Gamma(\xi)}$.*

Proof. It suffices to show that $\xi \in \overline{\Gamma(\zeta)}$ for any $\zeta \in \overline{\Gamma(\xi)}$. Since X is geodesically complete, there are geodesics $\sigma, \rho : \mathbb{R} \to X$ with $\sigma(\infty) = \xi$ and $\rho(\infty) = \zeta$. Let $\xi' = \sigma(-\infty)$ and $\zeta' = \rho(-\infty)$.

Since Γ satisfies the duality condition, ξ' is Γ-dual to ξ and ζ' is Γ-dual to ζ. Now ζ is Γ-dual to ξ' because the set of points Γ-dual to ξ' is Γ-invariant and closed. By Lemma 1.8 this implies that $\xi' \in \overline{\Gamma(\zeta')}$.

Choose a sequence (φ_n) in Γ with $\varphi_n(\zeta') \to \xi'$. For $m > 0$, let σ_m be a complete geodesic with $\sigma_m(0) = \sigma(m)$ and $\sigma_m(-\infty) = \varphi_n(\zeta')$, where $n = n(m) \geq m$ is chosen so large that $d(\sigma_m, \sigma(0)) \leq 1$. Then $\sigma_m(\infty) \to \xi$ and $\sigma_m(-\infty)$ is Γ-dual to $\sigma_m(\infty)$. Therefore $\sigma_m(\infty)$ is Γ-dual to $\zeta' = \varphi_n^{-1}(\sigma_m(-\infty))$. Hence ζ' is Γ-dual to $\xi = \lim_{m\to\infty} \sigma_m(\infty)$. Now Lemma 1.8 implies $\xi \in \overline{\Gamma(\zeta)}$. $\square$

1.10 Remark. The duality condition has other useful technical properties.

(1) Let X_1 and X_2 be Hadamard spaces with metrics d_1 and d_2 respectively and let X be the Hadamard space $X_1 \times X_2$ with the metric $d = \sqrt{d_1^2 + d_2^2}$. If Γ is a group of isometries of X satisfying the duality condition such that any $\varphi \in \Gamma$ is of the form $\varphi = (\varphi_1, \varphi_2)$, where φ_i is an isometry of X_i, $i = 1, 2$, then

$$\Gamma_1 := \{\varphi_1 \,|\, \text{there is an isometry } \varphi_2 \text{ of } X_2 \text{ with } (\varphi_1, \varphi_2) \in \Gamma\}$$

and the corresponding group Γ_2 satisfy the duality condition. This is clear since geodesics in X are pairs (σ_1, σ_2), where σ_1 is a geodesic of X_1 and σ_2 is a geodesic of X_2. See [Eb10,16] for applications of this property in the case of Hadamard manifolds.

Note that Γ_1 and/or Γ_2 need not act properly discontinuously even if Γ does. Thus a condition like cocompactness + proper discontinuity does not share the above property of the duality condition.

(2) Clearly the duality condition passes from a group Γ of isometries of a Hadamard space X to a larger group of isometries of X. Vice versa, if X is geodesically complete and separable and if Δ is a subgroup of finite index in Γ, then Δ satisfies the duality condition if Γ does [Eb10]. To see this, let $\sigma \in GX$ be a Γ-recurrent geodesic. Choose sequences (t_n) in $\mathbb{R}$ and (φ_n) in Γ such that $t_n \to \infty$ and $\varphi_n \circ g^{t_n}(\sigma) \to \sigma$. Now Δ has finite index in Γ. Hence we may assume, after passing to a subsequence, that $\varphi_n = \psi_n \varphi$, where $\varphi \in \Gamma$ is fixed and $\psi_n \in \Delta$ for all $n \in \mathbb{N}$. Then

$$\sigma_n' = \varphi_n \circ g^{t_n}(\sigma) \to \sigma$$

and

$$\sigma_{m,n} = (\psi_m \psi_n^{-1}) \circ g^{t_m - t_n}(\sigma_n') = \varphi_m \circ g^{t_m}(\sigma) \to \sigma \, .$$

Choose $m = m(n)$ convenient to get that $\sigma_{m(n),n} \to \sigma$. This shows that σ is nonwandering mod Δ. Now the set of geodesics in GX which are nonwandering mod Δ is closed, and the set of Γ-recurrent geodesics in GX is dense in GX, see Corollary 1.4. Hence Δ satisfies the duality condition.

We come back to the beginning of this section. We said there that we will be in need of a large group of isometries, and we made this precise by introducing the duality condition. One might think that the isometry group of a homogeneous Hadamard manifold X is large; however, it is not in our sense, except if X is a symmetric space (see [Eb10], Proposition 4.9).

1.11 Definition/Exercise/Question. Let X be a Hadamard space and Γ a group of isometries of X. The *limit set* $L(\Gamma) \subset X(\infty)$ of Γ is defined to be the set of all $\xi \in X(\infty)$ such that there is a sequence (φ_n) in Γ with $\lim_{n\to\infty} \varphi_n(x) = \xi$ for one (and hence any) $x \in X$. Show $L(\Gamma) = X(\infty)$

(i) if there is a bounded subset B in X with $\cup_{\varphi \in \Gamma} \varphi(B) = X$ or

(ii) if X is geodesically complete and Γ satisfies the duality condition.

In many results below we assume that X is geodesically complete and that Γ satisfies the duality condition. Can one (you) replace the duality condition by the assumption $L(\Gamma) = X(\infty)$.

2. Geodesic flows on Hadamard spaces

2.1 Definition. Let X be a geodesically complete Hadamard space and Γ a group of isometries of X. We say the geodesic flow of X is *topologically transitive* mod Γ if for any two open non-empty subsets U, V in GX there is $t \in \mathbb{R}$ and $\varphi \in \Gamma$ such that $g^t(U) \cap \varphi(V) \neq \emptyset$.

2.2 Remark. If X is separable, then the topology of X, and therefore the topology of GX, has a countable base. Hence the geodesic flow is topologically transitive mod Γ if and only if it has a *dense orbit* mod Γ: there is a σ in GX such that for any σ' in GX there are sequences (t_n) in $\mathbb{R}$ and φ_n in Γ with $\varphi_n g^{t_n}\sigma \to \sigma'$.

After all our preparations we are ready for the following result from [Eb3].

2.3 Theorem. *Let X be a geodesically complete, separable Hadamard space, and let Γ be a group of isometries of X satisfying the duality condition. Then the following are equivalent:*

(i) g^t is topologically transitive mod Γ;
(ii) for some $\xi \in X(\infty)$, the orbit $\Gamma(\xi)$ is dense in $X(\infty)$;
(iii) for every $\xi \in X(\infty)$, the orbit $\Gamma(\xi)$ is dense in $X(\infty)$;
(iv) $X(\infty)$ does not contain a non-empty Γ-invariant proper closed subset.

Proof. Clearly (iii) $\Leftrightarrow$ (iv) and (iii) $\Rightarrow$ (ii). The implication (i) $\Rightarrow$ (ii) is immediate from Remark 2.2 above. The implication (ii) $\Rightarrow$ (iii) follows from Proposition 1.9. It remains to prove (ii) $\Rightarrow$ (i).

We let $U(\infty)$ respectively $V(\infty)$ be the set of points $\sigma(\infty)$ in $X(\infty)$ with $\sigma \in U$ respectively V. Then $U(\infty)$ and $V(\infty)$ are open and non-empty. Applying (ii) we can assume $U(\infty) \subset V(\infty)$. Let $\sigma \in U$. Then there is a geodesic $\rho \in V$ with $\rho(\infty) = \sigma(\infty)$. Now Corollary 1.3 applies. $\square$

2.4 Theorem. *Let X be a geodesically complete and locally compact Hadamard space, and let Γ be a group of isometries of X satisfying the duality condition. Then the following are equivalent:*

(i) the geodesic flow (g^t) of X is not topologically transitive mod Γ;
(ii) $X(\infty)$ contains a non-empty Γ-invariant proper compact subset;
(iii) the geodesic flow (g^t) of X admits a non-constant Γ-invariant continuous first integral $f : GX \to \mathbb{R}$ such that $f(\sigma) = f(\rho)$ for all $\sigma, \rho \in GX$ with $\sigma(\infty) = \rho(\infty)$ or $\sigma(-\infty) = \rho(-\infty)$.

Each of these conditions implies that every geodesic of X is contained in a flat plane, that is, a convex subset of X isometric to the Euclidean plane.

Proof. X is separable because X is locally compact. Hence (i) $\Leftrightarrow$ (ii) by Theorem 2.3. The implication (iii) $\Rightarrow$ (ii) is clear since we assume the integral to be continuous, non-constant and Γ-invariant and since $X(\infty)$ is compact. It remains to show (ii) $\Rightarrow$ (iii) and the existence of the flat planes.

To that end, let $C \subset X(\infty)$ be a non-empty Γ-invariant proper compact subset. For $\sigma \in GX$ we set

$$f(\sigma) = \min_{\zeta \in C} \angle_{\sigma(0)}(\sigma(\infty), \zeta)\,. \tag{2.5}$$

We claim that the function f is an integral of the geodesic flow as asserted. Note that f is Γ-invariant since C is.

The continuity of angle measurement at a fixed point, see I.3.2, implies that for any $\sigma \in GX$ there is $\zeta \in C$ with $f(\sigma) = \angle_{\sigma(0)}(\sigma(\infty), \zeta)$. Obviously, or by Proposition II.4.2, we have that $f(g^t(\sigma))$ is monotonically not decreasing in t.

Suppose first that $\sigma \in GX$ is Γ-recurrent. Choose sequences (t_n) in $\mathbb{R}$ and (φ_n) in Γ such that $t_n \to \infty$ and $\varphi_n \circ g^{t_n}(\sigma) \to \sigma$. By the Γ-invariance of f and the monotonicity we have

$$f(\varphi_n \circ g^{t_n}(\sigma)) \;=\; f(g^{t_n}(\sigma)) \;\geq\; f(\sigma)\,.$$

The semicontinuity of angle measurement, see I.3.2, implies

$$\liminf_{n\to\infty} f(\varphi_n \circ g^{t_n}(\sigma)) \leq f(\sigma)\,.$$

We conclude that $f(g^t(\sigma) = f(\sigma)$ for all $t \geq 0$, and hence for all $t \in \mathbb{R}$ since $g^s(\sigma)$ is Γ-recurrent for all $s \in \mathbb{R}$. We also conclude that there is a point $\zeta(\sigma) \in C$ such that

$$f(g^t(\sigma)) \;=\; \angle_{\sigma(t)}(\sigma(\infty), \zeta(\sigma)) \;=\; \angle(\sigma(\infty), \zeta(\sigma)) \;=\; \min_{\zeta \in C} \angle(\sigma(\infty), \zeta)\,.$$

In particular, σ bounds a flat half plane H_σ with $\zeta(\sigma) \in H_\sigma(\infty)$ if $\sigma(\infty) \notin C$.

Now let $\sigma \in GX$ be arbitrary. Let (σ_n) be a sequence of Γ-recurrent unit speed geodesics converging to σ. Passing to a subsequence if necessary, the corresponding half planes H_{σ_n} and points $\zeta(\sigma_n)$ converge to a flat half plane H_σ along σ respectively a point $\zeta(\sigma) \in H_\sigma(\infty)$. We conclude

$$f(\sigma) \;=\; \lim_{n\to\infty} f(\sigma_n) \;=\; \angle(\sigma(\infty), \zeta(\sigma))\,.$$

Therefore f is continuous; now Lemma 1.1 implies that $f(\sigma) = f(\rho)$ if $\sigma(\infty) = \rho(\infty)$ respectively if $\sigma(-\infty) = \rho(-\infty)$; f is non-constant since C is proper.

Let $\sigma \in GX$ and let H_σ, $\zeta(\sigma)$ be as above. Let $\rho_t : \mathbb{R} \to X$ be a unit speed geodesic with $\rho_t(0) = \sigma(t)$ and $\rho_t(\infty) = \zeta(\sigma)$. For $s \in \mathbb{R}$, let $\sigma_s : \mathbb{R} \to X$ be a unit speed geodesic $\sigma_s(t) = \rho_t(s)$ and $\sigma_s(\infty) = \sigma(\infty)$. Then clearly

$$f(\sigma) \;=\; f(\sigma_s) \;=\; \angle_{\sigma_s(t)}(\zeta(\sigma), \sigma_s(\infty))\,.$$

Therefore $\rho_t|[s,\infty)$ and $\sigma_s|[t,\infty)$ span a flat cone $C_{s,t}$. This cone contains $\sigma([t,\infty))$ if $s < 0$. Now X is locally compact. Hence we obtain a flat plane containing σ as a limit of the cones C_{s_n,t_n} for some appropriate sequences $s_n \to -\infty$ and $t_n \to -\infty$. □

2.6 Problem. Suppose X is geodesically complete and irreducible with a group Γ of isometries satisfying the duality condition. Assume furthermore that every geodesic of X is contained in a flat plane. Claim: X is a symmetric space or a Euclidean building.

This is known in the smooth case, which excludes the case of Euclidean buildings. In Chapter IV we will discuss the proof. In the singular case, if $\dim X = n$ and every geodesic of X is contained in an n-flat, that is, a convex subset of X isometric to n-dimensional Euclidean space, then X is a Euclidean building [Kl].

3. The flat half plane condition

In this section we assume that X is a locally compact Hadamard space. Then, if $\sigma : \mathbb{R} \to X$ is a geodesic which does not bound a flat half plane, there is a constant $R > 0$ such that σ does not bound a flat strip of width R.

3.1 Lemma. *Let $\sigma : \mathbb{R} \to X$ be a unit speed geodesic which does not bound a flat strip of width $R > 0$. Then there are neighborhoods U of $\sigma(-\infty)$ and V of $\sigma(\infty)$ in $\overline{X}$ such that for any $\xi \in U$ and $\eta \in V$ there is a geodesic from ξ to η, and for any such geodesic σ' we have $d(\sigma', \sigma(0)) < R$. In particular, σ' does not bound a flat strip of width $2R$.*

Proof. If the assertion of the lemma does not hold, then there are sequences (x_n) and (y_n) in X with $x_n \to \sigma(-\infty)$ and $y_n \to \sigma(\infty)$ such that the unit speed geodesic segment σ_n from x_n to y_n, parameterized such that $\sigma_n(0)$ is the point on σ_n closest to $z := \sigma(0)$, satisfies $d(\sigma_n(0), z) \geq R$ for all $n \in \mathbb{N}$. Let z_n be the point on the geodesic from z to $\sigma_n(0)$ with $d(z, z_n) = R$. By the choice of $\sigma_n(0)$,

$$\angle_{\sigma_n(0)}(z, x_n),\ \angle_{\sigma_n(0)}(z, y_n) \geq \frac{\pi}{2}.$$

Hence also

$$\angle_{z_n}(z, x_n),\ \angle_{z_n}(z, y_n) \geq \frac{\pi}{2}.$$

Passing to a subsequence if necessary we assume $z_n \to z_\infty$. Then $d(z, z_\infty) = R$ and

$$\angle_{z_\infty}(z, \sigma(-\infty)),\ \angle_{z_\infty}(z, \sigma(\infty)) \ \geq\ \frac{\pi}{2}.$$

We also have

$$\angle_z(\sigma(-\infty), z_\infty) + \angle_z(\sigma(\infty), z_\infty) \ \geq\ \pi\,.$$

Applying Proposition I.5.8(i) we conclude

$$\angle_z(\sigma(-\infty), z_\infty) = \angle_{z_\infty}(\sigma(-\infty), z) = \angle_z(\sigma(\infty), z_\infty) = \angle_{z_\infty}(\sigma(\infty), z) = \frac{\pi}{2}\,.$$

Hence the unit speed rays ω_- respectively ω_+ from z_∞ to $\sigma(-\infty)$ and $\sigma(\infty)$ respectively span flat half strips with $\sigma|(-\infty, 0]$ respectively $\sigma|[0, \infty)$, see Proposition I.5.8(i). In particular, $\sigma(-t)$ respectively $\sigma(t)$ is the point on σ closest to $\omega_-(t)$ respectively $\omega_+(t)$, $t > 0$. Hence $d(\omega_-(t), \omega_+(t)) \geq 2t$, and hence ω_- and ω_+ combine to give a geodesic ω from $\sigma(-\infty)$ to $\sigma(\infty)$ through z_∞. Therefore ω is parallel to σ and ω and σ span a flat strip of width R. This is a contradiction. □

3.2 Lemma. *Let $\sigma : \mathbb{R} \to X$ be a unit speed geodesic which does not bound a flat half plane. Let (φ_n) be a sequence of isometries of X such that $\varphi_n(x) \to \sigma(\infty)$ and $\varphi_n^{-1}(x) \to \sigma(-\infty)$ for one (and hence any) $x \in X$. Then for n sufficiently large, φ_n has an axis σ_n such that $\sigma_n(\infty) \to \sigma(\infty)$, $\sigma_n(-\infty) \to \sigma(-\infty)$ as $n \to \infty$.*

Proof. There is an $R > 0$ such that σ does not bound a flat strip of width R. Let $x = \sigma(0)$ and choose $\varepsilon > 0, r > 0$ such that

$$U = U(x, \sigma(-\infty), r, \varepsilon) \text{ and } V = U(x, \sigma(\infty), r, \varepsilon).$$

are neighborhoods as in Lemma 3.1. By comparison with Euclidean geometry we find that there is an $R' > r$ such that for any $y \in X$ we have

$$(*) \qquad \begin{aligned} &d(\sigma(r), \sigma_{x,y}(r)) < \varepsilon \quad \text{or} \quad d(\sigma(-r), \sigma_{x,y}(r)) < \varepsilon \\ &\qquad\qquad \text{if } d(y,\sigma) \leq R \text{ and } d(y,x) \geq R'. \end{aligned}$$

By the definition of U and V this means that $y \in U \cup V$ if $d(y,\sigma) \leq R$ and $d(y,x) \geq R'$.

Assume that φ_n, for any n in a subsequence, fixes a point $x_n \in \overline{X}$ outside $U \cup V$ with $d(x,x_n) \geq 2R'$. Let $\sigma_n = \sigma_{x,x_n}$ be the unit speed geodesic (or ray respectively if $x_n \in X(\infty)$) from x to x_n. Then the displacement function $d(\varphi_n(z), z)$ is monotonically decreasing along σ_n. Let $y_n = \sigma_n(R')$ and $y_n' = \sigma_n(2R')$. Then

$$\begin{aligned} \angle_{y_n}(y_n', \varphi_n(y_n)) + \angle_{y_n}(y_n', \varphi_n^{-1}(y_n)) &= \\ \angle_{y_n}(y_n', \varphi_n(y_n)) + \angle_{\varphi_n(y_n)}(\varphi_n(y_n'), y_n) &\leq \pi \end{aligned}$$

since

$$d(y_n, \varphi_n(y_n)) = d(y_n, \varphi_n^{-1}(y_n)) \geq d(y_n', \varphi_n(y_n')) = d(y_n', \varphi_n^{-1}(y_n'))$$

(consider the quadrangle spanned by $y_n, y_n', \varphi_n(y_n), \varphi_n(y_n')$). Now

$$\angle_{y_n}(y_n', \varphi_n^{\pm 1}(y_n)) + \angle_{y_n}(x, \varphi_n^{\pm 1}(y_n)) \geq \angle_{y_n}(y_n', x) = \pi$$

and therefore

$$(**) \qquad \angle_{y_n}(x, \varphi_n(y_n)) + \angle_{y_n}(x, \varphi_n^{-1}(y_n')) \geq \pi.$$

We also have

$$d(\varphi_n^{\pm 1}(y_n), \varphi_n^{\pm 1}(x)) = d(y_n, x) = R',$$

and hence $\varphi_n(y_n) \to \sigma(\infty)$ and $\varphi_n^{-1}(y_n) \to \sigma(-\infty)$ (for the given subsequence). After passing to a further subsequence if necessary, we can assume $y_n \to y$. Then $d(y,\sigma) \geq R$ by $(*)$ since $y \notin U \cup V$ and $d(y,x) = R'$. By $(**)$, the rays from y to $\sigma(-\infty)$ respectively $\sigma(\infty)$ combine to give a geodesic parallel to σ. Therefore σ bounds a flat strip of width R, a contradiction. Hence we obtain that, for n sufficiently large, any fixed point of φ_n is contained in $U \cup V \cup \overline{B}(x, 2R')$.

Using Lemma 3.1 it is now easy to exclude that φ_n is elliptic or parabolic (for this compare Proposition II.3.4) for all n sufficiently large. For large n, the endpoints of an axis σ_n of φ_n have to be in $U \cup V$ by what we said above. It is clear from Lemma 3.1 that not both endpoints of σ_n are in one of the sets, U or V. Hence the lemma. $\square$

3.3 Lemma. *Let φ be an isometry of X with an axis $\sigma : \mathbb{R} \to X$, where σ is a unit speed geodesic which does not bound a flat half plane. Then*

(i) for any neighborhood U of $\sigma(-\infty)$ and any neighborhood V of $\sigma(\infty)$ in $\overline{X}$ there exists $N \in \mathbb{N}$ such that

$$\varphi^n(\overline{X} \setminus U) \subset V, \quad \varphi^{-n}(\overline{X} \setminus V) \subset U \quad \textit{for all } n \geq N\,;$$

(ii) for any $\xi \in X(\infty) \setminus \{\sigma(\infty)\}$ there is a geodesic σ_ξ from ξ to $\sigma(\infty)$, and any such geodesic does not bound a flat half plane. For $K \subset X(\infty) \setminus \{\sigma(\infty)\}$ compact, the set of these geodesics is compact (modulo parameterization).

Proof. Let $x = \sigma(0)$. We can assume that

$$U = U(x, \sigma(-\infty), R, \varepsilon) \text{ and } V = U(x, \sigma(\infty), R, \varepsilon).$$

By comparison with Euclidean geometry we find that

$$(*) \qquad \varphi^{-1}(U) \subset U \quad \text{and} \quad \varphi(V) \subset V\,.$$

If (i) is not true, there is a sequence $n_k \to \infty$ and, say, a sequence (x_k) in $\overline{X} \setminus U$ such that $\varphi^{n_k}(x_k) \notin V$. Passing to a subsequence, we may assume $x_k \to \xi \in \overline{X} \setminus U$. From $(*)$ we conclude $\varphi^m(x_k) \notin V$ for $0 \leq m \leq n_k$, and hence we get

$$(**) \qquad \varphi^m(\xi) \notin V \text{ for all } m \geq 0.$$

In particular, $\xi \in X(\infty)$. Now the set K of all $\xi \in \overline{X}(\infty) \setminus U$ satisfying $(**)$ is compact and invariant under φ^m, $m \geq 1$. By our assumption $K \neq \emptyset$. Now choose $\xi \in K$ such that $\angle_{\sigma(0)}(\sigma(\infty), \xi)$ is minimal. Since $\varphi(\sigma(\infty)) = \sigma(\infty)$,

$$\angle_{\sigma(0)}(\sigma(\infty), \varphi^m(\xi)) \leq \angle_{\sigma(ma)}(\sigma(\infty), \varphi^m(\xi)) = \angle_{\sigma(0)}(\sigma(\infty), \xi),$$

where $a > 0$ is the period of φ. By the choice of ξ we conclude

$$\angle_{\sigma(0)}(\sigma(\infty), \varphi^m(\xi)) = \angle_{\sigma(ma)}(\sigma(\infty), \varphi^m(\xi)),$$

hence the rays from $\sigma(0)$ and $\sigma(ma)$ to $\varphi^m(\xi)$ together with $\sigma|[0, ma]$ bound a flat convex region. Since m is arbitrary and since φ shifts σ, we conclude that σ bounds a flat half plane, a contradiction. Thus (i) follows. Now (ii) is an immediate consequence of (i) and Lemma 3.1. □

Let Γ be a group of isometries of the Hadamard space X. We say that a non-constant geodesic σ in X is Γ-*closed* if it is an axis of an axial isometry $\varphi \in \Gamma$.

3.4 Theorem. *Assume that X is a locally compact Hadamard space containing a unit speed geodesic $\sigma : \mathbb{R} \to X$ which does not bound a flat half plane and that Γ is a group of isometries of X satisfying the duality condition. Then we have:*

If $X(\infty)$ contains at least three points, then $X(\infty)$ is a perfect set: for any $\xi \in X(\infty)$, there is a sequence (ξ_n) in $X(\infty) \setminus \{\xi\}$ such that $\xi_n \to \xi$. Furthermore, for any two non-empty open subsets U, V of $X(\infty)$ there is a $\varphi \in \Gamma$ with

$$\varphi(X(\infty) \setminus U) \subset V,\ \varphi^{-1}(X(\infty) \setminus V) \subset U.$$

More precisely, there is a Γ-closed geodesic σ with $\sigma(-\infty) \in U$ and $\sigma(\infty) \in V$ such that σ does not bound a flat half plane.

Proof. By Lemma 3.2, there is an isometry φ_0 in Γ with an axis σ_0 which does not bound a flat half plane. Since $X(\infty)$ contains more than two points, there is a point $\xi \in X(\infty) \setminus \{\sigma_0(\pm\infty)\}$. By Lemma 3.3, $\varphi_0^{\pm n}(\xi) \to \sigma_0(\pm\infty)$, hence the first assertion holds for $\sigma_0(\pm\infty)$. Again by Lemma 3.3, there is a geodesic σ_1 from $\sigma_0(-\infty)$ to ξ and this geodesic does not bound a flat half plane. By Lemma 3.2, any two neighborhoods of the endpoints of σ_1 contain endpoints of an axis σ_1' of an isometry φ_1 in Γ and σ_1' does not bound a flat half plane. Then a sufficient high power of φ_1 maps $\sigma_0(+\infty)$ into the given neighborhood, hence the first assertion also holds for ξ.

If U, V are non-empty and open in $X(\infty)$, there is, by the first step, a point $\xi \in U - \sigma_0(\infty)$. By Lemma 3.3, there is a geodesic half plane. By Lemma 3.2 there is a Γ-closed geodesic σ_2 with $\sigma_2(-\infty) \in U$ such that σ_2 does not bound a flat half plane. Again by the first part, there is a point $\eta \in V - \sigma_2(-\infty)$ and, by Lemma 3.3, a geodesic σ_3 from $\sigma_2(-\infty)$ to η which does not bound a flat half plane. By Lemma 3.2, there is a Γ-closed geodesic σ as claimed. □

3.5 Theorem. *Let X and Γ be as in Theorem 3.4. Suppose $X(\infty)$ contains at least three points.*

(i) If X is geodesically complete, then the geodesic flow is topologically transitive mod Γ.

(ii) Γ contains free non-abelian subgroups.

Proof. (i) This is immediate from Theorem 2.4. Alternatively we can argue as follows: Theorem 3.4 and Lemma 3.3 imply that the Γ-orbit of any point $\xi \in X(\infty)$ is dense in $X(\infty)$. Hence the geodesic flow of X is topologically transitive mod Γ by Theorem 2.3.

(ii) By the first assertion of Theorem 3.4, $X(\infty)$ contains non-empty open subsets W_1, W_2 with

$$(*) \qquad W_1 \cap W_2 = \emptyset \quad \text{and} \quad W_1 \cup W_2 \neq X(\infty).$$

By the second assertion of Theorem 3.4 and Lemma 3.3, Γ contains isometries φ_1, φ_2 with axis σ_1, σ_2 which do not bound flat half planes such that $\sigma_i(\pm\infty) \in W_i$ and

$$(**) \qquad \varphi_i^{\pm k}(X(\infty) \setminus W_i) \subset W_i,\ i = 1, 2,$$

for all $k \neq 0$. Let $\xi \in X(\infty) \setminus (W_1 \cup W_2)$. It follows from (*) and (**) that a non-trivial word w in φ_1 and φ_2, considered as an element of Γ, maps ξ to W_i if φ_i is the first letter of w, $i = 1, 2$. Hence $w(\xi) \neq \xi$ and therefore $w \neq id$. It follows that φ_1 and φ_2 generate a free subgroup of Γ. □

3.6 Remark. If X is a Hadamard manifold with a unit speed geodesic $\sigma : \mathbb{R} \to X$ which does not bound a flat half plane, and if Γ is a group of isometries of X satisfying the duality condition, then the geodesic flow of X is topologically mixing mod Γ [Ba1]: for any two non-empty open subsets U, V in GX there is a $T \geq 0$ such that for any $t \geq T$ there is $\varphi \in \Gamma$ with $g^t(U) \cap \varphi(V) \neq \emptyset$. This does not hold in the general case we consider, even if we impose the strongest hyperbolicity assumption. For example, if X is the tree with all vertices of valence three and all edges of length one, and if U respectively V is the set of all $\sigma \in GX$ such that $\sigma(0)$ is ε-close to a vertex respectively the midpoint of an edge, where $\varepsilon < 1/4$, then $g^t(U) \cap \varphi(V) = \emptyset$ for all $t \in \mathbb{Z}$ and every isometry φ of X. It is, however, possible to strengthen the above result about topological transitivity and to obtain a property intermediate between it and topological mixing.

In Chapter IV we prove that a Hadamard manifold M with a group Γ of isometries satisfying the duality condition contains a unit speed geodesic which does not bound a flat half plane if and only if it contains a unit speed geodesic σ which does not bound an infinitesimal flat half plane. By the latter we mean that there is no parallel vector field V along σ such that $\|V\| = 1$, $V \perp \dot\sigma$ and $K(\dot\sigma \wedge V) \equiv 0$. It is easy to see that σ satisfies this condition if and only if σ is hyperbolic in the sense of the following definition.

3.7 Definition (experimental). Let X be a Hadamard space. We say that a unit speed geodesic $\sigma : \mathbb{R} \to X$ is *hyperbolic* if there are real numbers $a < b$ and constants $\varepsilon > 0$ and $\lambda < 1$ such that any geodesic segment $\gamma : [a, b] \to X$ with $\gamma(a) \in B_\varepsilon(\sigma(a))$ and $\gamma(b) \in B_\varepsilon(\sigma(b))$ satisfies

$$d(\gamma(\frac{a+b}{2}), \sigma) \leq \frac{\lambda}{2}(d(\gamma(a), \sigma) + d(\gamma(b), \sigma)).$$

It is unclear whether a geodesically complete Hadamard space X (with reasonable additional assumptions like local compactness and/or large group of isometries) contains a hyperbolic unit speed geodesic (in the above sense or a variation of it) if it contains a unit speed geodesic which does not bound a flat half plane.

Let X be a geodesically complete and locally compact Hadamard space. Then the set G_hX of hyperbolic geodesic is open in GX and invariant under isometries and the geodesic flow.

3.8 Corollary. *Let X be a geodesically complete and locally compact Hadamard space and Γ a group of isometries of X satisfying the duality condition. If X contains a hyperbolic geodesic, then the set of hyperbolic Γ-closed geodesics is dense in GX.*

Proof. By Lemma 3.2, any hyperbolic geodesic σ is the limit of a sequence (σ_n) of Γ-closed geodesics. Since G_hX is open in GX, σ_n is hyperbolic for n large. Now Theorem 3.5(i) implies that G_hX is dense in GX. $\square$

4. Harmonic functions and random walks on Γ

Let X be a locally compact Hadamard space containing a unit speed geodesic $\sigma : \mathbb{R} \to X$ which does not bound a flat half plane and Γ a countable group of isometries of X satisfying the duality condition. Assume furthermore that $X(\infty)$ contains at least three points.

Suppose μ is a probability measure on Γ whose support generates Γ as a semigroup. We say that a function $h : \Gamma \to \mathbb{R}$ is *μ-harmonic* if

$$h(\varphi) = \sum_{\psi \in \Gamma} h(\varphi\psi)\mu(\psi) \quad \text{for all } \varphi \in \Gamma\,. \tag{4.1}$$

Our objective is to show that Γ admits many μ-harmonic functions.

We define the probability measures $\mu^k, k \geq 0$, on Γ recursively by $\mu^0 := \delta$, the Dirac measure at the neutral element of Γ, and

$$\mu^k(\varphi) = \sum_{\psi \in \Gamma} \mu^{k-1}(\psi)\mu(\psi^{-1}\varphi) \quad k \geq 1\,. \tag{4.2}$$

Then $\mu^1 = \mu$. An easy computation shows that $h : \Gamma \to \mathbb{R}$ is μ^k-harmonic if h is μ-harmonic. Now the support of μ generates Γ as a semigroup and hence for any $\varphi \in \Gamma$ there is a $k \geq 1$ such that $\mu^k(\varphi) > 0$.

For $k, l \geq 0$ we also have $\mu^k * \mu^l = \mu^{k+l}$, where

$$\mu^k * \mu^l(\varphi) = \sum_{\psi \in \Gamma} \mu^k(\psi)\mu^l(\psi^{-1}\varphi)\,. \tag{4.3}$$

For a probability measure ν on $X(\infty)$ and $k \geq 0$ define the *convolution* $\mu^k * \nu$ by

$$\int_{X(\infty)} f(\xi) d(\mu^k * \nu)(\xi) = \sum_{\Gamma} \int_{X(\infty)} f(\varphi\xi) d\nu(\xi) \mu^k(\varphi)\,, \tag{4.4}$$

where f is a bounded measurable function on $X(\infty)$. Then

$$\mu^k * (\mu^l * \nu) = \mu^{k+l} * \nu\,, \quad k, l \geq 0\,. \tag{4.5}$$

Since the space of probability measures on $X(\infty)$ is weakly compact, the sequence

$$\frac{1}{n+1}(\nu + \mu * \nu + \mu * \mu * \nu + \ldots + (\mu *)^n \nu)$$

has weakly convergent subsequences, and a weak limit $\bar{\nu}$ will be *stationary* (with respect to μ), that is, $\mu * \bar{\nu} = \bar{\nu}$. Thus the set of stationary probability measures is not empty.

Now let ν be a fixed stationary probability measure. By (4.5) we have

$$\mu^k * \nu = \nu \quad \text{for all } k \geq 0. \tag{4.6}$$

If there is a point $\xi \in X(\infty)$ such that $\nu(\xi) > 0$, then there is also a point $\xi_0 \in X(\infty)$ such that $\nu(\xi_0)$ is maximal since ν is a finite measure. For $k \geq 1$, the definition of $\mu^k * \nu$ implies

$$\nu(\xi_0) = \sum_\Gamma \nu(\varphi^{-1}\xi_0)\mu^k(\varphi),$$

and thus $\nu(\varphi^{-1}\xi_0) = \nu(\xi_0)$ for all φ in the support of μ^k by the maximality of $\nu(\xi_0)$. But for any $\varphi \in \Gamma$ there is $k \geq 1$ such that $\mu^k(\varphi) > 0$. Hence $\nu(\varphi^{-1}\xi_0) = \nu(\xi_0)$ for all $\varphi \in \Gamma$. But this is absurd since the orbit of ξ_0 under Γ is infinite. Hence

$$\nu \text{ is not supported on points.} \tag{4.7}$$

Thus for any $\xi \in X(\infty)$ and $\varepsilon > 0$ there is a neighbourhood U of ξ such that $\nu(U) < \varepsilon$.

Let f be a bounded measurable function on $X(\infty)$. Define a function h_f on Γ by

$$h_f(\varphi) = \int_{X(\infty)} f(\varphi\xi)d\nu(\xi) = \int_{X(\infty)} f(\xi)d(\varphi\nu)(\xi). \tag{4.8}$$

Since we have $\mu * \nu = \nu$ we obtain

$$\sum_\Gamma h_f(\varphi\psi)\mu(\psi) = \sum_\Gamma \int_{X(\infty)} f(\varphi\psi\xi)d\nu(\xi)\mu(\psi) = \int_{X(\infty)} f(\varphi\xi)d\nu(\xi) = h_f(\varphi),$$

and therefore h_f is a μ-harmonic function. We will show now that $h_f(\varphi_n) \to f(\xi)$ if ξ is a point of continuity of f and if $\varphi_n(x) \to \xi$ for one (and hence any) $x \in X$.

4.9 Lemma. *Let $x_0 \in X$, $R > 0$ and $\varepsilon > 0$ be given. Let $(\varphi_n) \subset \Gamma$ be a sequence such that $\varphi_n x_0 \to \xi \in X(\infty)$. Then there is an open subset $U \subset X(\infty)$ with $\nu(U) < \varepsilon$ and a $\varphi \in \Gamma$ such that for some subsequence (φ_{n_k}) of (φ_n) we have $\varphi_{n_k}\varphi(X \setminus U) \subset U(x_0, \xi, R, \varepsilon)$.*

Proof. After passing to a subsequence if necessary we may assume $\varphi_n^{-1}x_0 \to \eta \in X(\infty)$. By Theorem 3.4 there is a $\varphi_0 \in \Gamma$ with an axis σ which does not bound a flat half plane and with open neighborhoods U^- of $\sigma(-\infty)$ and U^+ of $\sigma(\infty)$ in $\overline{X}$ such that

$$\eta \in \overline{X} \setminus U^+ \text{ and } U^- \cap U(x_0, \xi, R, \varepsilon) = \emptyset.$$

By increasing R and diminishing ε and U^- if necessary we can assume $x_0 \notin U^-$ and $\nu(U^-) < \varepsilon$. By Lemmas 3.1 and 3.3 there is a neighborhood V^- of $\sigma(-\infty)$ in U^- such that for any two points $y, z \in \overline{X} \setminus U^-$ and any $x \in V^-$ the geodesics $\sigma_{x,y}$ and $\sigma_{x,z}$ exists and $d(\sigma_{x,y}(t), \sigma_{x,z}(t)) < \varepsilon/2$ for $0 \le t \le R$.

Now choose N so large that

$$\varphi_0^N(\overline{X} \setminus V^-) \subset U^+, \varphi_0^{-N}(\overline{X} \setminus U^+) \subset V^-$$

and let $\varphi = \varphi_0^N$. If n is large, then $\varphi_n^{-1} x_0 \in \overline{X} \setminus U^+$ since $\varphi_n^{-1} x_0 \to \eta$. Then $x = \varphi^{-1}\varphi_n^{-1} x_0 \in V^-$ and for $y \in \overline{X} \setminus U^-$ arbitrary we get, for $0 \le t \le R$,

$$\begin{aligned} &d(\sigma_{x_0,\xi}(t), \sigma_{x_0,\varphi_n\varphi(y)}(t)) \\ &\quad \le d(\sigma_{x_0,\xi}(t), \sigma_{x_0,\varphi_n\varphi(x_0)}(t)) + d(\sigma_{x_0,\varphi_n\varphi(x_0)}(t), \sigma_{x_0,\varphi_n\varphi(y)}(t)) \,. \end{aligned}$$

The first term on the right hand side tends to 0 as n tends to ∞ since $\varphi_n(\varphi x_0) \to \xi$. The second term is less than $\varepsilon/2$ for n sufficiently large since $x_0, y \in \overline{X} \setminus U^-$. Hence $\varphi_n\varphi(\overline{X} \setminus U^-) \subset U(x_0, \xi, R, \varepsilon)$ for all n sufficiently large. □

4.10 Theorem (Dirichlet problem at infinity). *Let X be a locally compact Hadamard space containing a unit speed geodesic $\sigma : \mathbb{R} \to X$ which does not bound a flat half plane. Assume that $X(\infty)$ contains at least 3 points. Suppose that Γ is a countable group of of isometries of X satisfying the duality condition and that μ is a probability measure on Γ whose support generates Γ as a semigroup.*

Then we have: if f is a bounded measurable function on $X(\infty)$ and $\xi \in X(\infty)$ is a point of continuity for f, then h_f is continuous at ξ; that is, if $(\varphi_n) \subset \Gamma$ is a sequence such that $\varphi_n x \to \xi$ for one (and hence any) $x \in X$, then $h_f(\varphi_n) \to f(\xi)$. In particular, if $f : X(\infty) \to \mathbb{R}$ is continuous, then h_f is the unique μ-harmonic function on Γ extending continuously to f at infinity.

Proof. Without loss of generality we can assume $f(\xi) = 0$. If there is a sequence $(\varphi_n) \subset \Gamma$ such that $\varphi_n x_0 \to \xi$ and such that $h_f(\varphi_n)$ does not tend to $f(\xi)$, then there is such a sequence (φ_n) with the property that $\limsup_{n\to\infty} |h_f(\varphi_n)|$ is maximal among all these sequences. We fix such a sequence and set

$$\delta = \limsup_{n\to\infty} |h_f(\varphi_n)|.$$

By passing to a subsequence if necessary we can assume that the lim sup is a true limit. Since ξ is a point of continuity of f, there are $R > 0$ and $\varepsilon > 0$ such that

$$|f(\eta)| < \delta/3 \text{ for all } \eta \in X(\infty) \cap U(x_0, \xi, R, \varepsilon).$$

By Lemma 4.9 there is a $\varphi \in \Gamma$ and an open subset U of $X(\infty)$ such that $\nu(U) < \varepsilon$ and $\varphi_{n_k}\varphi(X(\infty) \setminus U) \subset U(x_0, \xi, R, \varepsilon)$ for a subsequence (φ_{n_k}) of (φ_n). Then

$$\begin{aligned} |h_f(\varphi_{n_k}\varphi)| &\le \left| \int_{X(\infty)\setminus U} f(\varphi_{n_k}\varphi\eta) d\nu(\eta) \right| + \left| \int_U f(\varphi_{n_k}\varphi\eta) d\nu(\eta) \right| \\ &\le \nu(X(\infty) \setminus U) \cdot \delta/3 + \nu(U) \sup |f|. \end{aligned}$$

Hence $|h_f(\varphi_{n_k}\varphi)| \leq \delta/2$ for all k if we choose $\varepsilon > 0$ small enough.

Now there is a $k \geq 1$ such that $\mu^k(\varphi) = \alpha > 0$. We split Γ into three disjoint parts G, L and $\{\varphi\}$, where G is finite and $\mu^k(L)\sup|f| < \alpha\delta/2$. Since h_f is μ^k-harmonic we get

$$\begin{aligned}\limsup_{k\to\infty} |h_f(\varphi_{n_k})| &= \limsup_{k\to\infty}\left|\sum_{\Gamma} h_f(\varphi_{n_k}\psi)\mu^k(\psi)\right| \\ \leq \limsup_{k\to\infty}\left|\sum_{G} h_f(\varphi_{n_k}\psi)\mu^k(\psi)\right| &+ \limsup_{k\to\infty}|h_f(\varphi_{n_k}\varphi)|\alpha + \alpha\delta/2 \\ < \mu(G)\cdot\delta + \alpha\delta/2 + \alpha\delta/2 &\leq (1-\alpha)\delta + \alpha\delta = \delta.\end{aligned}$$

This is a contradiction. □

We now consider the left-invariant random walk on Γ defined by μ; that is, the transition probability from $\varphi \in \Gamma$ to $\psi \in \Gamma$ is given by $\mu(\varphi^{-1}\psi)$. For $\psi_1, \dots, \psi_k \in \Gamma$, the probability P that a sequence (φ_n) satisfies $\varphi_i = \psi_i, 1 \leq i \leq k$, is by definition equal to

$$\mu(\psi_1)\mu(\psi_1^{-1}\psi_2)\cdots\mu(\psi_{k-1}^{-1}\psi_k).$$

Since the support of μ generates Γ and since Γ is not amenable (Γ contains a free subgroup), random walk on Γ is *transient*, see [Fu3, p.212], that is, $d(x, \varphi_n x) \to \infty$ for P-almost any sequence $(\varphi_n) \subset \Gamma$. An immediate consequence of Theorem 4.10 and the Martingale Convergence Theorem is the following result.

4.11 Theorem. *Let X, Γ and μ be as in Theorem 4.10. Then for almost any sequence (φ_n) in Γ, the sequence $(\varphi_n x), x \in X$, tends to a limit in $X(\infty)$. The hitting probability is given by ν. In particular, the stationary measure ν is uniquely determined.*

Proof. If h is a bounded μ-harmonic function, then $M_k((\varphi_n)) := h(\varphi_k)$, $k \geq 1$, defines a martingale and hence $M_k((\varphi_n))$ converges P-almost surely by the Martingale Convergence Theorem, see [Fu3].

Let $\xi \neq \eta \in X(\infty)$ and let U, V be neighborhoods of U and V in $X(\infty)$. Then there is a continuous function f on $X(\infty)$ which is negative at ξ, positive at η and 0 outside $U \cup V$. Applying the Martingale Convergence Theorem in the case $h = h_f$ we conclude that the set of sequences (φ_n) in Γ such that $(\varphi_n(x))$, $x \in X$, has accumulation points close to both ξ and η has P-measure 0. Now $X(\infty)$ is compact, hence the first assertion.

Let $f : X(\infty) \to \mathbb{R}$ be continuous and set $M_k((\varphi_n)) = h_f(\varphi_k)$, $k \geq 1$. If π denotes the hitting probability at $X(\infty)$, then

$$\int f d\pi = \int (\lim M_k) dP.$$

On the other hand, since (M_k) is a martingale,

$$\begin{aligned}\int (\lim M_k) dP &= \int M_k dP \\ &= h_f(e) = \int f d\nu\end{aligned}$$

Hence $\nu = \pi$ and the proof is complete. $\square$

We may ask whether $(X(\infty), \nu)$ is the *Poisson boundary* of (Γ, μ). By that we mean that the assertions of Theorem 4.11 hold and that every bounded μ-harmonic function h on Γ is given as $h = h_f$, where f is an appropriate bounded measurable function on $X(\infty)$. This is true if X, Γ and μ satisfy the assumptions of Theorem 4.10 and

(i) Γ is finitely generated and μ has finite first moment with respect to the word norm of a finite system of generators of Γ;

(ii) Γ is properly discontinuous and cocompact on X.

For the proof (in the smooth case) we refer to [BaL1]; see alsox [Kai3, Theorem 3] for a simplification of the argument in [BaL1]. We omit a more elaborate discussion of the Poisson boundary since the relation between cocompactness on the one hand and the duality condition on the other is unclear (as of now).

For the convenience of the reader, the Bibliography contains many references in which topics related to our random walks are discussed: Martin boundary, Brownian motion, potential theory ... for Hadamard spaces and groups acting isometrically on them.

CHAPTER IV

Rank Rigidity

In this chapter we prove the Rank Rigidity Theorem. The proof proper is in Sections 4–7, whereas Sections 1–3 are of a preliminary nature. The initiated geometer should skip these first sections and start with Section 4. In Section 1 we discuss geodesic flows on Riemannian manifolds, in Section 2 estimates on Jacobi fields in terms of sectional curvature (Rauch comparison theorem) and in Section 3 the regularity of Busemann functions.

1. Preliminaries on geodesic flows

Let M be a smooth n-dimensional manifold with a connection D. Let TM be the tangent bundle of M and denote by $\pi : TM \to M$ the projection. Recall that the charts of M define an atlas for TM in a natural way, turning TM into a smooth manifold of dimension $2n$. Namely, if $x : U \to U' \subset \mathbb{R}^n$ is a chart of M and

$$X_i = \frac{\partial}{\partial x^i}, \quad 1 \leq i \leq n,$$

are the basic vector fields over U determined by x, then we define

$$\hat{x} : \pi^{-1}(U) \to U' \times \mathbb{R}^n; \quad \hat{x}(v) = (x(p), dx\,|_p(v)) \,, \tag{1.1}$$

where $p = \pi(v)$. If we write v as a linear combination of the $X_i(p)$,

$$v = \sum \xi^i X_i(p) \,,$$

then

$$(\xi^1, \dots, \xi^n) =: \xi(v) = dx\,|_p(v) \,,$$

and hence

$$\hat{x} = (\bar{x}, \xi) \,,$$

where $\bar{x} = x \circ \pi$. It is easy to see that the family of maps $\hat{x}$, where x is a chart of M, is a C^∞-atlas for TM.

The connection D allows for a convenient description of TTM, the tangent bundle of the tangent bundle. To that end, we introduce the vector bundle $\mathcal{E} \to TM$,

$$\mathcal{E} = \pi^*(TM) \oplus \pi^*(TM).$$

The fibre $\mathcal{E}_v$ of $\mathcal{E}$ at $v \in TM$ is

$$\mathcal{E}_v = T_pM \oplus T_pM\,, \tag{1.2}$$

where $p = \pi(v)$ is the foot point of v. We define a map $\mathcal{I} : TTM \to \mathcal{E}$ as follows:

$$\mathcal{I}(Z) = \left(\dot{c}(0), \frac{DV}{dt}(0)\right), \tag{1.3}$$

where V is a smooth curve in TM with $\dot{V}(0) = Z$ and $c = \pi \circ V$. To see that (1.3) defines a map $TTM \to \mathcal{E}$, we express it in a local coordinate chart $\hat{x}$ as above. Then $\hat{x} \circ V = (\sigma, \xi)$, where $\sigma = x \circ c$ and $V = \sum \xi^i X_i$. Hence

$$\left(\dot{c}(0), \frac{DV}{dt}(0)\right)$$

is represented by

$$\left(\dot{\sigma}(0), \dot{\xi}(0) + \Gamma(\dot{\sigma}(0), \xi(0))\right)\,, \tag{1.4}$$

where Γ denotes the *Christoffel symbol* with respect to the chart x and where we use $\hat{x} \oplus \hat{x}$ as a trivialization for $\mathcal{E}$. We conclude that $\mathcal{I}$ is well-defined, smooth and linear on the fibres. It is easy to see that $\mathcal{I}$ is surjective on the fibres. Now $\mathcal{I}$ is bijective since the fibres of TTM and $\mathcal{E}$ have the same dimension $2n$. Therefore:

1.5 Lemma. *The map $\mathcal{I} : TTM \to \mathcal{E}$ is a vector bundle isomorphism.* □

1.6 Exercise. For a coordinate chart x and a tangent vector $v = \sum \xi^k X_k(p)$ as in the beginning of this section show that

$$\mathcal{I}\left(\left.\frac{\partial}{\partial \xi_i}\right|_v\right) = (0, X_i(p))$$

and

$$\mathcal{I}\left(\left.\frac{\partial}{\partial \bar{x}_i}\right|_v\right) = (X_i(p), D_{X_i}V(p))\,,$$

where $V = \sum \xi^k X_k$ is the "constant extension" of v to a vector field on the domain of the coordinate chart x.

Henceforth we will identify TTM and $\mathcal{E}$ via $\mathcal{I}$ without further reference to the identification map $\mathcal{I}$.

For $v \in TTM$ and $p = \pi(v)$, the splitting

$$T_vTM = T_pM \oplus T_pM$$

singles out two n-dimensional subspaces. The first one is the *vertical space*

$$\mathcal{V}_v = \ker \pi_{*v} = \{(0, Y) \mid Y \in T_pM\}. \tag{1.7}$$

The *vertical distribution* $\mathcal{V}$ is smooth and integrable. The integral manifolds of $\mathcal{V}$ are the fibres of π, that is, the tangent spaces of M. Here we visualize the tangent spaces of M as vertical to the manifold – a convenient picture when thinking of TM as a bundle over M. The second n-dimensional subspace is the *horizontal space*

$$\mathcal{H}_v = \{(X, 0) \mid X \in T_pM\}. \tag{1.8}$$

The *horizontal distribution* $\mathcal{H}$ of TM is smooth but, in general, not integrable. By definition, a curve V in TM is *horizontal*, that is, $\dot{V}$ belongs to $\mathcal{H}$, if and only if $V' \equiv 0$, where V' denotes the covariant derivative of V along $\pi \circ V$; in other words, V is horizontal if and only if V is parallel along $c = \pi \circ V$.

The map, which associates to two smooth horizontal vector fields $\mathcal{X}, \mathcal{Y}$ on TM the vertical component of $[\mathcal{X}, \mathcal{Y}]$, is bilinear and, in each variable, linear over the smooth functions on TM. Hence the vertical component of $[\mathcal{X}, \mathcal{Y}]$ at $v \in TM$ only depends on $\mathcal{X}(v)$ and $\mathcal{Y}(v)$. To measure the non-integrability of $\mathcal{H}$, it is therefore sufficient to consider only special horizontal vector fields.

Now let X be a smooth vector field on M. The *horizontal lift* $X^{\mathcal{H}}$ of X is defined by

$$X^{\mathcal{H}}(v) = (X(\pi(v)), 0), \quad v \in TM. \tag{1.9}$$

By definition, X is *π-related* to $X^{\mathcal{H}}$, that is, $\pi_* \circ X^{\mathcal{H}} = X \circ \pi$. This allows us to express the Lie bracket of horizontal lifts of vector fields X, Y on M in a simple way.

1.10 Lemma. *Let X, Y be smooth vector fields on M. Then*

$$[X^{\mathcal{H}}, Y^{\mathcal{H}}](v) = ([X, Y]_p, \; -R(X_p, Y_p)v),$$

where $p = \pi(v)$ and the index p indicates evaluation at p.

Proof. Since X and Y are π-related to $X^{\mathcal{H}}$ and $Y^{\mathcal{H}}$ respectively, we have

$$\pi_{*v}([X^{\mathcal{H}}, Y^{\mathcal{H}}](v)) = [X, Y]_p.$$

This shows the claim about the horizontal component.

Since the vertical component only depends on X_p and Y_p, it suffices to consider the case that $[X, Y] = 0$ in a neighborhood of p. Denote by φ_t the flow of X and by ψ_s the flow of Y. Then the flows Φ_t of $X^{\mathcal{H}}$ and Ψ_s of $Y^{\mathcal{H}}$ are parallel translations along φ_t and ψ_s respectively. Let $v \in TM$ and let

$$V(t) = \Psi_{-t}\Phi_{-t}\Psi_t\Phi_t(v).$$

Since $[X, Y] = 0$ in a neighborhood of p, $V(t) \in T_pM$ for all t sufficiently small and

$$[X^{\mathcal{H}}, Y^{\mathcal{H}}](v) = \lim_{t \to 0} \frac{1}{t^2}(V(t) - v).$$

The RHS is precisely the formula for $R(Y_p, X_p)v$ using parallel translation. □

1.11 Exercise. Show that $[X^{\mathcal{H}}, \mathcal{Z}] = 0$, where $X^{\mathcal{H}}$ is as above and $\mathcal{Z}$ is a smooth vertical field of TM.

Recall that the *geodesic flow* (g^t) acts on TM by

$$g^t(v) = \dot{\gamma}_v(t), \quad t \in \mathcal{D}_v, \tag{1.12}$$

where $\mathcal{D}_v$ denotes the maximal interval of definition for the geodesic γ_v determined by $\dot{\gamma}_v(0) = v$. By definition, the connection is *complete* if $\mathcal{D}_v = \mathbb{R}$ for all $v \in TM$. To compute the differential of the geodesic flow, let $v \in TM$ and $(X, Y) \in T_vTM$. Represent (X, Y) by a smooth curve V through v, that is, $\dot{c}(0) = X$ for $c = \pi \circ V$ and $\dfrac{DV}{dt}(0) = Y$. Let J be the Jacobi field of the geodesic variation $\gamma_{V(s)}$ of γ_v. By the definition of the splitting of TTM we have

$$\begin{aligned} g^t_{*v}(X, Y) &= \frac{d}{ds} g^t(V(s))\Big|_{s=0} \\ &= \left(\frac{\partial}{\partial s}\gamma_{V(s)}(t), \frac{D}{\partial s}\dot{\gamma}_{V(s)}(t)\right)\Big|_{s=0} \\ &= \left(J(t), \frac{D}{\partial t}\frac{\partial}{\partial s}\gamma_{V(s)}(t)\right)\Big|_{s=0} \\ &= (J(t), J'(t)), \end{aligned}$$

where the prime indicates covariant differentiation along γ_v. Hence:

1.13 Proposition. *The differential of the geodesic flow is given by*

$$g^t_{*v}(X, Y) = (J(t), J'(t)),$$

where J is the Jacobi field along γ_v with $J(0) = X$ and $J'(0) = Y$. □

We now turn our attention to the case that D is the Levi-Civita connection of a Riemannian metric on M. By using the corresponding splitting of TTM, we obtain a Riemannian metric on TM, the *Sasaki metric*,

$$< (X_1, Y_1), (X_2, Y_2) > := < X_1, X_2 > + < Y_1, Y_2 > . \tag{1.14}$$

There is also a natural 1-form α on TM,

$$\alpha_v((X,Y)) = < v, X > . \tag{1.15}$$

Using Lemma 1.10 it is easy to see that the differential $\omega = d\alpha$ is given by

$$\omega((X_1,Y_1),(X_2,Y_2)) = < X_2, Y_1 > - < X_1, Y_2 > . \tag{1.16}$$

The Jacobi equation tells us that ω is invariant under the geodesic flow. The 1-form α is not invariant under the geodesic flow on TM.

1.17 Exercise. Compute the expressions for α and ω with respect to local coordinates $\hat{x} = (\bar{x}, \xi)$ as in the beginning of this section.

Note that ω is a nondegenerate closed 2-form and hence a *symplectic form* on TM. The measure on TM defined by $|\omega^n|$ is proportional to the volume form of the Sasaki metric. It is called the *Liouville measure.* The Liouville measure is invariant under the geodesic flow since ω is.

Denote by $SM \subset TM$ the unit tangent bundle. We have

$$T_vSM = \{(X,Y) \mid X, Y \in T_pM,\, Y \perp v\}, \tag{1.18}$$

where $v \in SM$ and $p = \pi(v)$. The geodesic flow leaves SM invariant. The restrictions of α and ω to SM, also denoted by α and ω respectively, are invariant under the geodesic flow since $< \dot{\gamma}_v, J > \equiv \text{const}$ if J is a Jacobi field along γ_v with $J' \perp \dot{\gamma}_v$. The $(2n-1)$-form $\alpha \wedge \omega^{n-1}$ is proportional to the volume form of the Sasaki metric (1.14) on SM. In particular, α is a *contact form.* The measure on SM defined by $\alpha \wedge \omega^{n-1}$ is also called the *Liouville measure.* It is invariant under the geodesic flow on SM since α and $\omega = d\alpha$ are.

A first and rough estimate of the differential of the geodesic flow with respect to the Sasaki metric is as follows, see [BBB1].

1.19 Proposition. *For $v \in SM$ set $R_vX = R(X,v)v$. Then we have*

$$\|g^t_*(v)\| \le \exp\left(\frac{1}{2}\int_0^t \|id - R_{g^s(v)}\| ds\right) .$$

Proof. Let J be a Jacobi field along γ_v. Then

$$\begin{aligned}
(\langle J, J\rangle + \langle J', J'\rangle)' &= 2\langle J', J + J''\rangle \\
&= 2\left\langle J', \left(id - R_{g^s(v)}\right) J\right\rangle \\
&\le 2\|J\| \cdot \|J'\| \cdot \|id - R_{g^s(v)}\| \\
&\le (\langle J, J\rangle + \langle J', J'\rangle)\, \|id - R_{g^s(v)}\|
\end{aligned}$$

□

We finish this section with some remarks on the geometry of the unit tangent bundle.

1.20 Proposition [Sas]. *A smooth curve V in SM is a geodesic in SM if and only if*

$$V'' = -< V', V' > V \quad \text{and} \quad \dot{c}' = R(V', V)\dot{c},$$

where $c = \pi \circ V$ and prime denotes covariant differentiation along c.

Proof. The energy of a piecewise smooth curve $V : [a, b] \to SM$ is given by

$$E(V) = \frac{1}{2}\int_a^b \|\dot{c}\|^2 dt + \frac{1}{2}\int_a^b \|V'\|^2 dt.$$

The second integral is the energy of the curve $\tilde{V}$ in the unit sphere at $c(0)$ obtained by parallel translating the vectors $V(t)$ along c to $V(0)$.

Assume V is a critical point of E. By considering variations of V which leave fixed the curve c of foot points and $V(a)$ and $V(b)$, we see that $\tilde{V}$ is a geodesic segment in the unit sphere. Hence V satisfies the first equation.

Now consider an arbitrary variation $V(s, \cdot)$ of V with foot point $c(s, \cdot)$ which keeps $V(\cdot, a)$ and $V(\cdot, b)$ fixed. Then

$$\begin{aligned} 0 = \frac{dE}{ds} &= \int_a^b \left[\left\langle \frac{D}{ds}\dot{c}, \dot{c}\right\rangle + \left\langle \frac{D}{ds}V', V'\right\rangle\right] dt \\ &= \int_a^b \left[\left\langle \frac{D}{dt}\frac{\partial c}{\partial s}, \dot{c}\right\rangle + \left\langle \frac{D}{dt}\frac{D}{ds}V, V'\right\rangle + \left\langle R\left(\frac{\partial c}{\partial s}, \frac{\partial c}{\partial t}\right)V, V'\right\rangle\right] dt \\ &= \int_a^b \left[\frac{d}{dt}\left(\left\langle \frac{\partial c}{\partial s}, \dot{c}\right\rangle\right) - \left\langle \frac{\partial c}{\partial s}, \dot{c}'\right\rangle + \frac{d}{dt}\left(\left\langle \frac{D}{ds}V, V'\right\rangle\right)\right. \\ &\qquad \left. - \left\langle \frac{D}{ds}V, V''\right\rangle + \left\langle R(V', V)\dot{c}, \frac{\partial c}{\partial s}\right\rangle\right] dt. \end{aligned}$$

Note that

$$\int_a^b \frac{d}{dt}\left\langle \frac{\partial c}{\partial s}, \dot{c}\right\rangle dt = 0 = \int_a^b \frac{d}{dt}\left\langle \frac{D}{ds}V, V'\right\rangle dt$$

since $c(s, a), c(s, b), V(s, a)$, and $V(s, b)$ are kept fixed. Also $\langle (D/ds)V, V''\rangle = 0$ since V'' and V are collinear by the first equation and $\langle (D/ds)V, V\rangle = 0$ because V is a curve of unit vectors. Hence the second equation is satisfied.

Conversely, our computations show that V is a critical point of the energy if it satisfies the equations in Proposition 1.20. □

1.21 Remarks. (a) The equations in Proposition 1.20 imply

$$\langle V', V'\rangle' = 0 = \langle \dot{c}, \dot{c}\rangle'$$

and therefore

$$\|V'\| = \text{const.} \quad \text{and} \quad \|\dot{c}\| = \text{const.}$$

(b) The great circle arcs in the unit spheres S_pM, $p \in M$, are geodesics in SM. In particular, we have that the Riemannian submersion $\pi : SM \to M$ has totally geodesic fibres.

(c) If V is horizontal, then V is a geodesic if and only if $c = \pi \circ V$ is a geodesic in M.

1.22 Exercise. Let M be a surface and V a geodesic in the unit tangent bundle of M. Assume that V is neither horizontal nor vertical. Then $V' \neq 0$ and $\dot{c} \neq 0$, where $c = \pi \circ V$, and

$$\begin{aligned} V''(t) &= -\|V'\|^2 \cdot V(t) \\ k(t) &= \frac{\|V'\|}{\|\dot{c}\|} K(c(t)) \end{aligned}$$

where k is the geodesic curvature of c with respect to the normal n such that $(\dot{c}, n)$ and (V, V') have the same orientation and K is the Gauss curvature of M. Use this to discuss geodesics in the unit tangent bundles of surfaces of constant Gauss curvature.

2. Jacobi fields and curvature

In Proposition 1.13 we computed the differential of the geodesic flow (g^t) on TM in terms of the canonical splitting of TTM. We obtained that

$$g^t_{*v}(X, Y) = (J(t), J'(t)), \tag{2.1}$$

where J is the Jacobi field along the geodesic γ_v with $J(0) = X$ and $J'(0) = Y$. In order to study the qualitative behaviour of (g^t) it is therefore useful to get estimates on Jacobi fields.

2.2 Lemma. *Let $\gamma : \mathbb{R} \to M$ be a unit speed geodesic, and suppose that the sectional curvature of M along γ is bounded from above by a constant κ. If J is a Jacobi field along γ which is perpendicular to $\dot{\gamma}$, then $\|J\|''(t) \geq -k\|J\|(t)$ for all t with $J(t) \neq 0$.*

Proof. The straightforward computation does the job:

$$\begin{aligned} \|J\|'' &= \left(\frac{< J', J >}{\|J\|}\right)' = \frac{1}{\|J\|^2}\left(< J'', J > \|J\| + < J', J' > \|J\| - \frac{< J', J >^2}{\|J\|}\right) \\ &= \frac{1}{\|J\|^3}\left(- < R(J, \dot{\gamma})\dot{\gamma}, J > \|J\|^2 + \|J'\|^2\|J\|^2 - < J', J >^2\right) \geq -k\|J\|. \end{aligned}$$

□

The same computation gives the following result.

2.3 Lemma. *Let $\gamma : \mathbb{R} \to M$ be a unit speed geodesic and suppose that the sectional curvature of M along γ is nonpositive. If J is a Jacobi field along γ (not necessarily perpendicular to $\dot{\gamma}$), then $\|J(t)\|$ is convex as a function of t. In particular, there are no conjugate points along γ.* □

For $\kappa \in \mathbb{R}$ denote by sn_κ respectively cs_κ the solution of

$$j'' + \kappa j = 0 \tag{2.4}$$

with $\mathrm{sn}_\kappa(0) = 0$, $\mathrm{sn}'_\kappa(0) = 1$ respectively $\mathrm{cs}_\kappa(0) = 1$, $\mathrm{cs}'_\kappa(0) = 1$. Then $\mathrm{sn}'_\kappa = \mathrm{cs}_\kappa$ and $\mathrm{cs}'_\kappa = -\kappa\,\mathrm{sn}_\kappa$. We also set $\mathrm{tg}_\kappa = \dfrac{\mathrm{sn}_\kappa}{\mathrm{cs}_\kappa}$ and $\mathrm{ct}_\kappa = \dfrac{\mathrm{cs}_\kappa}{\mathrm{sn}_\kappa}$. Lemma 2.2 implies the following version of the Rauch Comparison Theorem.

2.4 Proposition. *Let $\gamma : \mathbb{R} \to M$ be a unit speed geodesic and suppose that the sectional curvature of M along γ is bounded from above by a constant κ. If J is a Jacobi field along γ with $J(0) = 0$, $J'(0) \perp \dot\gamma(0)$ and $\|J'(0)\| = 1$, then*

$$\|J(t)\| \;\geq\; \mathrm{sn}_\kappa(t) \quad \textit{and} \quad \|J\|'(t) \;\geq\; \mathrm{ct}_\kappa(t)\|J(t)\|$$

for $0 < t < \pi/\sqrt{\kappa}$.

Proof. For any Jacobi field J along γ and perpendicular to $\dot\gamma$ and any solution j of (2.4) we have by Lemma 2.2

$$(\|J\|'j - \|J\|j')' \;\geq\; 0$$

wherever j and $\|J\|$ are positive. For J as in Proposition 2.4,

$$\lim_{t\to 0+} \frac{\|J\|(t)}{\mathrm{sn}_\kappa(t)} = \lim_{t\to 0+} \frac{\|J\|'(t)}{\mathrm{sn}'_\kappa(t)} = 1$$

by l'Hospital's rule. Hence

$$\|J\|'(t) \;\geq\; \mathrm{ct}_\kappa(t)\|J(t)\| \quad \text{and} \quad \|J(t)\| \geq \mathrm{sn}_\kappa(t)$$

for t small and then, by continuation, for all t as claimed. □

The proof of the estimates for Jacobi fields in the case of a lower bound on the curvature is different.

2.5 Proposition. *Let $\gamma : \mathbb{R} \to M$ be a unit speed geodesic and suppose that the sectional curvature of M along γ is bounded from below by a constant λ. If J is a Jacobi field along γ with $J(0) = 0$, $J'(0) \perp \dot\gamma(0)$ and $\|J'(0)\| = 1$, then*

$$\|J(t)\| \;\leq\; \mathrm{sn}_\lambda(t) \quad \textit{and} \quad \|J'(t)\| \;\leq\; \mathrm{ct}_\lambda(t)\|J(t)\|$$

if there is no pair of conjugate points along $\gamma \mid [0,t]$.

For the proof of this version of the Rauch Comparison Theorem we refer to [Kar], [BuKa].

In the rest of this section we deal with manifolds of nonpositive or negative curvature. The infinitesimal version of the Flat Strip Theorem I.5.8(ii) is as follows.

2.6 Lemma. *Let M be a manifold of nonpsitive curvature and $\gamma : \mathbb{R} \to M$ a unit speed geodesic. Let J be a Jacobi field along γ. Then the following conditions are equivalent:*
(i) J is parallel; (ii) $\|J(t)\|$ is constant on $\mathbb{R}$; (iii) $\|J(t)\|$ is bounded on $\mathbb{R}$.
Each of these condition implies that $< R(J,\dot\gamma)\dot\gamma, J > \equiv 0$.

Proof. Clearly (i) ⇒ (ii) ⇒ (iii). Now (iii) ⇒ (ii) since $\|J(t)\|$ is convex in t, see Lemma 2.3. If $\|J(t)\|$ is constant, then

$$0 \;= < J, J >'' =\; 2\,(< J', J' > - < R(J,\dot\gamma)\dot\gamma, J >)\,.$$

and hence $J'(t) \equiv 0$ since the sectional curvature is nonpositive. Hence (ii) ⇒ (i). □

2.7 Definition. Let M be a manifold of nonpositive curvature and $\gamma : \mathbb{R} \to M$ a unit speed geodesic. We say that a Jacobi field J along γ is *stable* if $\|J(t)\| \leq C$ for all $t \geq 0$ and some constant $C \geq 0$.

2.8 Proposition. *Let M be a manifold of nonpositive curvature and $\gamma : \mathbb{R} \to M$ a unit speed geodesic. Set $p = \gamma(0)$.*

(i) For any $X \in T_pM$, there exists a unique stable Jacobi field J_X along γ with $J_X(0) = X$.

(ii) Let (γ_n) be a sequence of unit speed geodesics in M with $\gamma_n \to \gamma$. For each n, let J_n be a Jacobi field along γ_n. Assume that $J_n(0) \to X \in T_pM$ and that $\|J_n(t_n)\| \leq C$, where $t_n \to \infty$ and C is independent of n. Then $J_n \to J_X$ and $J_n' \to J_X'$.

Proof. Uniqueness in (i) follows from the convexity of $\|J(t)\|$. Existence follows from (ii) since we may take $\gamma_n = \gamma$ and J_n the Jacobi field with $J_n(0) = X$, $J_n(n) = 0$ (recall that γ has no conjugate points). Now let γ_n, J_n, t_n and C be as in (ii). Fix $m \geq 0$. By convexity we have

$$\|J_n(t)\| \leq C, \quad 0 \leq t \leq m,$$

for all n with $t_n \geq m$. By a diagonal argument we conclude that any subsequence (n_k) has a further subsequence such that the corresponding subsequence of (J_n) converges to a Jacobi field J along γ with $\|J(t)\| \leq C$ for all $t \geq 0$. Now the uniqueness in (i) implies $J = J_X$ and (ii) follows. □

2.9 Proposition. *Let M be a manifold of nonpositive curvature, $\gamma : \mathbb{R} \to M$ a unit speed geodesic and J a stable Jacobi field along γ perpendicular to $\dot\gamma$.*

(i) If the curvature of M along γ is bounded from above by $\kappa = -a^2 \leq 0$, then

$$\|J(t)\| \leq \|J(0)\|e^{-at} \quad \text{and} \quad \|J'(t)\| \geq a\|J(t)\| \quad \text{for all } t \geq 0.$$

(ii) If the curvature of M along γ is bounded from below by $\lambda = -b^2 \leq 0$, then

$$\|J(t)\| \geq \|J(0)\|e^{-bt} \quad \text{and} \quad \|J'(t)\| \leq b\|J(t)\| \quad \text{for all } t \geq 0.$$

Proof. Let J_n be the Jacobi field along γ with $J_n(0) = J(0)$ and $J_n(n) = 0$, $n \geq 1$. Then $J_n \to J$ and $J_n' \to J'$ by the previous lemma. If $\kappa = -a^2 \leq 0$ is an upper bound for the curvature along γ, then by Proposition 2.4

$$\frac{\|J_n(t)\|}{\|J_n(0)\|} \leq \frac{\sinh(a(n-t))}{\sinh(an)}$$

and

$$\|J_n\|'(t) \leq -a\coth(a(n-t))\|J_n(t)\|, \quad 0 < t < n,$$

where we have $\mathrm{sn}_\kappa(s) = \frac{1}{a}\sinh(as)$ ($:= s$ if $a = 0$) and $\mathrm{ct}_\kappa(s) = a\coth(as)$ ($:= 1$ if $a = 0$). Hence (i). In the same way we conclude (ii) from Proposition 2.5. □

The following estimate will be needed in the proof of the absolute continuity of the stable and unstable foliations in the Appendix on the ergodicity of geodesic flows.

Proposition 2.10. *Let M be a Hadamard manifold and assume that the sectional K_M of M is bounded by $-b^2 \leq K_M \leq -a^2 < 0$. Denote by d_S the distance in the unit tangent bundle SM of M.*

Then for every constant $D > 0$ there exist constants $C = C(a,b) \geq 1$ and $T = T(a,b) \geq 1$ such that

$$d_S(g^t v, g^t w) \leq Ce^{-at} d_S(v,w), \quad 0 \leq t \leq R,$$

where v, w are inward unit vectors to a geodesic sphere of radius $R \geq T$ in M with foot points x, y of distance $d(x,y) \leq D$

Proof. Since the curvature of M is uniformly bounded, the metric

$$d_1(v,w) = d(\gamma_v(0), \gamma_w(0)) + d(\gamma_v(1), \gamma_w(1))$$

is equivalent to d_S and therefore it suffices to consider d_1. For the same reason, there is a constant $C_1 > 0$ such that the interior distance of any pair of points x, y on any geodesic sphere in M is bounded by $C_1 d(x,y)$ as long as $d(x,y) \leq D$. Now the interior distance of $\gamma_v(t)$ and $\gamma_w(t)$ in the sphere of radius $R - t$ about $p = \gamma_v(R) = \gamma_w(R)$ can be estimated by a geodesic variation γ_s, $0 \leq s \leq 1$, such that $\gamma_0 = \gamma_v$, $\gamma_1 = \gamma_w$, $\gamma_s(R) = p$ and $\gamma_s(0) \in S_p(R)$, $0 \leq s \leq 1$. Arguing as in the proof of Proposition 2.9, we conclude that there is a constant C_2 such that the asserted inequality holds for $0 \leq t \leq R - 1$ (recall that we are using the distance d_1). Now the differential of the geodesic flow g^τ, $0 \leq \tau \leq 1$, is uniformly bounded by a constant $C_3 = C_3(b)$, see Proposition 1.19. Hence $C = C_2 C_3 e^a$ is a constant as desired. □

3. Busemann functions and horospheres

We come back to the description of Busemann functions in Exercise II.2.6.

3.1 Proposition. *Let M be a Hadamard manifold and let $p \in M$. Then a function $f : M \to \mathbb{R}$ is a Busemann function based at p if and only if (i) $f(p) = 0$; (ii) f is convex; (iii) f has Lipschitz constant 1; (iv) for any $q \in M$ there is a point $q_1 \in M$ with $f(q) - f(q_1) = 1$.*

Proof. Let $f : M \to \mathbb{R}$ be a function satisfying the conditions (i)–(iv). Let $q \in M$ be arbitrary. Set $q_0 := q$ and define $q_n, n \geq 1$, recursively to be a point in M with $d(q_n, q_{n-1}) = 1$ and $f(q_n) - f(q_{n-1}) = -1$. Let $\sigma_n : [n-1, n] \to M$ be the unit speed geodesic segment from q_{n-1} to $q_n, n \geq 1$. Since f has Lipschitz constant 1, the concatenation

$$\sigma_q := \sigma_1 * \sigma_2 * \ldots : [0, \infty) \to M$$

is a unit speed ray with

$$f(\sigma_q(t)) = f(q) - t, \quad t \geq 0.$$

We show now that all the rays σ_q, $q \in M$, are asymptotic to σ_p. We argue by contradiction and assume that $\sigma_q(\infty) \neq \sigma_p(\infty)$. Let σ be the unit speed ray from p with $\sigma(\infty) = \sigma_q(\infty)$. Then $d(\sigma(t), \sigma_p(t)) \to \infty$ and there is a first $t_1 > 0$ such that $d(\sigma(t_1), \sigma_p(t_1)) = 1$. For $t \geq t_1$ we have

$$d^2(\sigma(t), \sigma_p(t)) \geq 2t^2(1 - \sin\alpha), \quad \alpha = \frac{1}{2t_1},$$

by comparison with Euclidean geometry. Hence the midpoint m_t between $\sigma(t)$ and $\sigma_p(t)$ satisfies

$$(*) \qquad d(m_t, p) \leq t\cos\alpha, \quad t \geq t_1.$$

Now f is convex and Lipschitz, hence

$$\begin{aligned} f(m_t) &\leq \frac{1}{2}(f(\sigma(t) + f(\sigma_p(t))) \\ &\leq \frac{1}{2}\big(f(\sigma_q(t)) + f(\sigma_p(t)) + d(p,q)\big) \leq -t + C, \end{aligned}$$

where $C = (d(p,q) + f(q))/2$. This contradicts $(*)$.

It follows easily that f is a Busemann function. The other direction is clear. □

The characterization in Proposition 3.1 implies that Busemann functions are C^1, cf. [BGS, p.24]. Using that Busemann functions are limits of normalized distance functions, we actually get that they are C^2. For $\xi \in M(\infty)$ and $p \in M$ let $\sigma_{p,\xi}$ be the unit speed ray from p to ξ and set $v_\xi(p) = \dot\sigma_{p,\xi}(0)$.

3.2 Proposition (Eberlein, see [HeIH]). *Let M be a Hadamard manifold, $p \in M$ and $\xi \in M(\infty)$. If $f = b(\xi, p, .)$ is the Busemann function at ξ based at p, then f is C^2 and we have*

$$\operatorname{grad} f = -v_\xi \quad \textit{and} \quad D_X \operatorname{grad} f(q) = -J_X'(0)\ , q \in M\ , X \in T_qM\,,$$

where J_X is the stable Jacobi field along $\sigma_{q,\xi}$ with $J_X(0) = X$.

Proof. Let (p_n) be a sequence in M converging to ξ and let

$$f_n(q) = b(p_n, p, q) = d(q, p_n) - d(q, p).$$

Then $f_n \to f$ uniformly on compact subsets of M. The gradient of f_n on $M \setminus \{p_n\}$ is the negative of the smooth field of unit vectors v_n pointing at p_n. Now

$$\|v_n - v_\xi\|(q) \leq \angle_q(p_n, \xi).$$

By comparison with Euclidean geometry, the right hand side tends to 0 uniformly on compact subsets of M as $n \to \infty$. Hence f is C^1 and $\operatorname{grad} f = -v_\xi$.

Now let X be a smooth vector field on M. Then

$$D_{X(q)} \operatorname{grad} f_n = -J_n'(q,0),$$

where $J_n(q,.)$ is the Jacobi field along σ_{q,p_n} with $J_n(0) = X(q)$ and $J_n(q,t_n) = 0$, where $t_n = d(q,p_n)$. By Proposition 2.8, $J_n'(q,0)$ tends to $J_{X(q)}'(0)$ uniformly on compact subsets of M. Hence f is C^2 and the covariant derivative of $\operatorname{grad} f$ is as asserted. □

3.3 Remark. In general, Busemann functions are not three times differentiable, even if the metric of M is analytic, see [BBB2]. However, if the first k derivatives of the curvature tensor R are uniformly bounded on M and the sectional curvature of M is pinched between two negative constants, then Busemann functions are C^{k+2}.

For $v \in SM$ set $v(\infty) := \gamma_v(\infty)$. We define

$$W^{so}(v) = \{w \in SM \mid w(\infty) = v(\infty)\}.$$

For v fixed and $\xi = v(\infty)$,

$$W^{so}(v) = \{-\operatorname{grad} f(q) \mid q \in M\}.$$

where f is a Busemann function at ξ. Proposition 3.2 implies that $W^{so}(v)$ is a C^1-submanifold of SM. The submanifolds $W^{so}(v)$, $v \in SM$, are a continuous foliation of SM. By Proposition 3.2, the tangent distribution of W^{so} is

$$E^{so}(v) = \{(X,Y) \mid Y = J_X'(0)\},$$

where J_X is the stable Jacobi field along γ_v with $J_X(0) = X$. The regularity of the foliation W^{so} is a very intricate question.

Remark 3.4. The following is known in the case that M is the universal cover (in the Riemannian sense) of a compact manifold of *strictly* negative curvature:

(a) For each $v \in SM$, $W^{so}(v)$ is the weak stable manifold of v with respect to the geodesic flow and hence smooth. (Compare also Remark 3.3.)

(b) The foliation W^{so} is a smooth foliation of SM if and only if M is a symmetric space (of rank one), see [Gh, BFL, BCG].

(c) If the dimension of M is 2 or if the sectional curvature of M is strictly negatively $\frac{1}{4}$-pinched, then the distribution E^{so} is C^1, see [Ho1, HiPu]. If the dimension of M is 2 and E^{so} is C^2, then it is C^∞, see [HurK].

(d) The foliation W^{so} is Hölder and absolutely continuous, see [Ano1], [AnSi] and the Appendix below.

4. Rank, regular vectors and flats

As before, we let M be a Hadamard manifold and denote by SM the unit tangent bundle of M. For a vector $v \in SM$, the *rank* of v is the dimension of the vector space $\mathcal{J}^p(v)$ of parallel Jacobi fields along γ_v and the *rank* of M is the minimum of $\operatorname{rank}(v)$ over $v \in SM$.

Clearly we have $\operatorname{rank}(w) \leq \operatorname{rank}(v)$ for all vectors in a sufficiently small neighborhood of a given vector $v \in SM$. We define $\mathcal{R}$, the set of *regular vectors* in SM, to be those vectors v in SM such that $\operatorname{rank}(w) = \operatorname{rank}(v)$ for all w sufficiently close to v. The set of regular vectors of rank m is denoted by $\mathcal{R}_m$. The sets $\mathcal{R}$ and $\mathcal{R}_m$ are open. Moreover, $\mathcal{R}$ is dense in SM since the rank function is semicontinuous and integer valued. If $k = \operatorname{rank}(M)$, then the set $\mathcal{R}_k$ is nonempty and consists of all vectors in SM of rank k.

For $v \in SM$ we can identify T_vTM with the space $\mathcal{J}(v)$ of Jacobi fields along γ_v by associating to $(X, Y) \in T_vTM$ the Jacobi field J along γ with $J(0) = X$, $J'(0) = Y$. We let $\mathcal{F}(v)$ be the subspace of T_vSM corresponding to the space $\mathcal{J}^p(v)$ of parallel Jacobi fields along γ_v. That is, $\mathcal{F}(v)$ consists of all $(X, 0)$ such that the stable Jacobi field J_X along γ_v determined by $J_X(0) = X$ is parallel. The distribution $\mathcal{F}$ is tangent to SM and invariant under isometries and the geodesic flow of M. Note that $\mathcal{F}$ has constant rank locally precisely at vectors in $\mathcal{R}$. On each nonempty set $\mathcal{R}_m$, $m \geq \operatorname{rank}(M)$, $\mathcal{F}$ has constant rank m, and we shall see that $\mathcal{F}$ is smooth and integrable on $\mathcal{R}_m$. Moreover, for each vector $v \in \mathcal{R}_m$ the integral manifold of $\mathcal{F}$ through v contains an open neighborhood of v in the set $\mathcal{P}(v)$ of all vectors in SM that are parallel to v.

4.1 Lemma. *For every integer $m \geq \operatorname{rank}(M)$, the distribution $\mathcal{F}$ is smooth on $\mathcal{R}_m$.*

Proof. For each vector $v \in \mathcal{R}_m$, we consider the symmetric bilinear form

$$Q_T^v(X, Y) = \int_{-T}^{T} < R(X, \dot{\gamma}_v)\dot{\gamma}_v, Y > dt,$$

where X, Y are arbitrary parallel vector fields (not necessarily Jacobi) along γ_v. Here R denotes the curvature tensor of M, and T is a positive number. Since the linear transformation $w \to R(w, \dot{\gamma}_v)\dot{\gamma}_v$ is symmetric and negative semidefinite, a parallel vector field X on γ_v is in the nullspace of Q_T^v if and only if for all $t \in [-T, T]$

$$R(X(t), \dot{\gamma}_v(t))\dot{\gamma}_v(t) = 0.$$

Hence such a vector field X is a Jacobi field on $\gamma_v[-T, T]$. It follows that for a small neighborhood U of v in $\mathcal{R}_m$ and a sufficiently large number T the nullspace of Q_T^w is precisely $\mathcal{J}^p(w)$ for all $w \in U$. For a fixed T the form Q_T^w depends smoothly on w. Since the dimension of the nullspace is constant in U, $\mathcal{F}(w)$ also depends smoothly on w. $\square$

We now integrate parallel Jacobi fields to produce flat strips in M.

4.2 Lemma. *The distribution $\mathcal{F}$ is integrable on each nonempty set $\mathcal{R}_m \subset SM$, $m \geq \operatorname{rank}(M)$. The maximal arc-connected integral manifold through $v \in \mathcal{R}_m$ is an open subset of $\mathcal{P}(v)$. In particular, if $w \in \mathcal{P}(v) \cap \mathcal{R}_m$, then $\mathcal{P}(v)$ and $P(v)$ are smooth m-dimensional manifolds near w and $\pi(w)$ respectively.*

Proof. To prove the first assertion we begin with the following obeservation: let $\sigma : (-\varepsilon, \varepsilon) \to \mathcal{R}_m$ be a smooth curve tangent to $\mathcal{F}$. Then $\sigma(s)$ is parallel to $\sigma(0) = v$ for all s. To verify this we consider the geodesic variation $\gamma_s(t) = (\pi \circ g^t)(\sigma(s))$ of $\gamma_0 = \gamma_v$. The variation vector fields Y_s of (γ_s) are given by

$$Y_s(t) = (d\pi \circ dg^t)(\frac{d}{ds}\sigma(s)).$$

The vector fields Y_s are parallel Jacobi fields along γ_s since $(d/ds)\sigma(s)$ belongs to $\mathcal{F}$. For each t the curve $\alpha_t : u \to \gamma_u(t)$ has velocity $(d/du)\alpha_t(u) = Y_u(t)$, and hence, since each Y_u is parallel, the lengths of the curves $\alpha_t(I)$ are constant in t for any interval $I \subset (-\varepsilon, \varepsilon)$. Therefore, for every s the convex function $t \to d(\gamma_0(t), \gamma_s(t))$ is bounded on $\mathbb{R}$ and hence constant. This implies that the geodesics γ_s are parallel to γ_0 and $\sigma(s) = \dot{\gamma}_s(0)$ is parallel to $\sigma(0) = v$.

Now let $v \in \mathcal{R}_m$ be given, and let $\mathcal{X}, \mathcal{Y}$ be smooth vector fields defined in a neighborhood of v in $\mathcal{R}_m$ that are tangent to $\mathcal{F}$. Denote by ϕ_ξ^s and ϕ_η^s the flows of $\mathcal{X}$ and $\mathcal{Y}$ respectively. The discussion above shows that if $w \in \mathcal{R}_m$ is parallel to v then so are $\phi_\xi^s(w)$ and $\phi_\eta^s(w)$ for every small s. Hence $\sigma(s^2) = \phi_\eta^{-s} \circ \phi_\xi^{-s} \circ \phi_\eta^s \circ \phi_\xi^s(v)$ is parallel to v for every small s. Therefore, $[\mathcal{X}, \mathcal{Y}](v) = (d/ds)\sigma(s)|_{s=0}$ belongs to $\mathcal{F}$, and it follows that $\mathcal{F}$ is integrable on $\mathcal{R}_m$.

The third assertion of the lemma follows from the second, which we now prove. If Q denotes the maximal arc-connected integral manifold of $\mathcal{F}$ through v, then $Q \subset \mathcal{P}(v)$ by the discussion above. Now let $w \in Q$ be given, and let $O \subset \mathcal{R}_m$ be a normal coordinate neighborhood of w relative to the metric in SM. We complete the proof by showing that

$$O \cap \mathcal{P}(w) = O \cap \mathcal{P}(v) \subset Q.$$

If $w' \in O \cap \mathcal{P}(v)$ is distinct from w, let $\sigma(s)$ be the unique geodesic in M from $\sigma(0) = \pi(w)$ to $\sigma(1) = \pi(w')$. The parallel field $w(s)$ along $\sigma(s)$ with $w(0) = w$ has values in $\mathcal{P}(v)$ since either $\gamma_w = \gamma_{w'}$ or γ_w and $\gamma_{w'}$ bound a flat strip in M. Hence $w[0,1]$ is the shortest geodesic in SM from $w(0) = w$ to $w(1) = w'$ and must therefore lie in $O \subset \mathcal{R}_m$. Since the curve $w(s)$, $0 \leq s \leq 1$, lies in $\mathcal{P}(v) \cap \mathcal{R}_m$ it is tangent to $\mathcal{F}$ and hence $w' \in w[0,1] \subset Q$. ☐

4.3 Lemma. *Let N^* be a totally geodesic submanifold of a Riemannian manifold N. Let γ be a geodesic of N that lies in N^*, and let J be a Jacobi field in N along γ. Let J_1 and J_2 denote the components of J tangent and orthogonal to N^* respectively. Then J_1 and J_2 are Jacobi fields in N along γ.*

Proof. Let R denote the curvature tensor of N, and let v be a vector tangent to N^* at a point p. Since N^* is totally geodesic it follows that

$$R(X, v)v \in T_pN^* \quad \text{if } X \in T_pN^*$$

and hence, since $Y \to R(Y,v)v$ is a symmetric linear operator, that

$$R(X,v)v \in (T_pN^*)^\perp \quad \text{if } X \in (T_pN^*)^\perp .$$

Since J_1'' is tangent and J_2'' perpendicular to N^* we see that

$$J_1'' + R(J_1, \dot\gamma)\dot\gamma$$

is the component of

$$0 = J'' + R(J, \dot\gamma)\dot\gamma$$

tangent to N^*, and hence must be zero. $\square$

4.4 Proposition. *Suppose the isometry group of M satisfies the duality condition. Then for every vector $v \in \mathcal{R}_m$, the set $P(v)$ is an m-flat, that is, an isometrically and totally geodesically embedded m-dimensional Euclidean space.*

Proof. Let $v \in \mathcal{R}_m$ be given. By Lemma 4.2 for any $v' \in \mathcal{R}_m$ the sets $\mathcal{P}(v') \cap \mathcal{R}_m$ and $P(v')$ are smooth m-manifolds near v' and $\pi(v')$ respectively, and $\pi : \mathcal{P}(v') \cap \mathcal{R}_m \to P(v')$ is an isometry onto an open neighborhood of $\pi(v')$ in $P(v')$.

We can choose a neighborhood $U \subset \mathcal{R}_m$ of v and a number $\varepsilon > 0$ such that for every $v' \in U$ the exponential map at $\pi(v')$ is a diffeomorphism of the open ε-ball in $T_{\pi(v')}P(v')$ onto its image $D_\varepsilon(v') \subset P(v')$.

Let $q \in P(v)$ be given. We assert that $q \in D_\varepsilon \subset P(v)$, where D_ε is the diffeomorphic image under the exponential map at q of the open ε-ball in some m-dimensional subspace of T_qM. Let $w = v(q)$ be the vector at q parallel to v. By Corollary III.1.3 we can choose sequences $s_n \to \infty$, $v_n \to v$ and (ϕ_n) in Γ such that $(d\phi_n \circ g^{s_n})(v_n) \to w$ as $n \to +\infty$. For sufficiently large n, $v_n \in U$ and

$$\phi_n(D_\varepsilon(v_n)) = D_\varepsilon(d\phi_n(v_n)) \subset P(d\phi_n(v_n)).$$

If $D_\varepsilon(v') \subset P(v')$ for some $v' \subset \mathcal{R}$, then clearly $D_\varepsilon(g^s v') \subset P(v') = P(g^s v')$ for any real number s since $P(v')$ is the union of flat strips containing $\gamma_{v'}$. Hence

$$D_\varepsilon(g^{s_n} \circ d\phi_n(v_n)) = D_\varepsilon(d\phi_n \circ g^{s_n}(v_n)) \subset P(d\phi_n \circ g^{s_n}(v_n))$$

for sufficiently large n. Therefore, the open m-dimensional ε-balls $D_\varepsilon(d\phi_n \circ g^{s_n}(v_n))$ converge to an open m-dimensional ε-ball $D_\varepsilon \subset P(w) = P(v)$ as asserted.

We show first that $D_\varepsilon = B_\varepsilon(q) \cap P(v)$, where $B_\varepsilon(q)$ is the open ε-ball in M with center q. Since $q \in P(v)$ was chosen arbitrary this will show that $P(v)$ is an m-dimensional submanifold without boundary of M, and it will follow that $P(v)$ is complete and totally geodesic since $P(v)$ is already closed and convex. Suppose that D_ε is a proper subset of $B_\varepsilon(q) \cap P(v)$. Choose a point $q' \in B_\varepsilon(q) \cap P(v)$ such that the geodesic from q to q' is not tangent to D_ε at q. The set $P(v)$ contains the convex hull D' of D_ε and q' and hence contains a subset $D'' \subset D'$ that is diffeomorphic to an open $(m+1)$-dimensional ball. Similarly $P(v)$ contains the convex hull of D'' and $p = \pi(v)$, but this contradicts the fact that $P(v)$ has dimension m at p.

We show next that $P(v)$ has zero sectional curvature in the induced metric. By a limit argument we can choose a neighborhood O of v in $T_{\pi(v)}P(v)$ such that $O \subset \mathcal{R}_m$ and for any $w \in O$ the geodesic γ_w admits no nonzero parallel Jacobi field that is orthogonal to $P(v)$. Let $w \in O$ be given, and let Y be any parallel Jacobi field along γ_w. If Z is the component of Y orthogonal to $P(v)$, then Z is a Jacobi field on γ_w by Lemma 4.3, and $\|Z(t)\| \leq \|Y(t)\|$ is a bounded convex function on $\mathbb{R}$. By Lemma 2.6, Z is parallel on γ_w and hence identically zero since $w \in O$. Hence

$$T_{\pi(w)}P(w) = \operatorname{span}\{Y(0) : Y \in \mathcal{J}^p(w)\} \subset T_{\pi(v)}P(v).$$

Since $P(v)$ and $P(w)$ are totally geodesic submanifolds of M of the same dimension m it follows that $P(v) = P(w)$ for all $w \in O$. The vector field $p \to w(p)$ is globally parallel on $P(w) = P(v)$. Since this is true for any $w \in O$ we obtain m linearly independent globally parallel vector fields in $P(v)$. Hence $P(v)$ is flat. $\square$

5. An invariant set at infinity

In this section we discuss Eberlein's modification of the Angle Lemma in [BBE], see [EbHe]. We will also use the notation P_v for $P(v)$. Since the distribution $\mathcal{F}$ is smooth on $\mathcal{R}$, we have the following conclusion of Proposition 4.4.

5.1 Lemma. *The function $(v,p) \mapsto d(p, P_v)$ is continuous on $\mathcal{R} \times M$.* $\square$

5.2 Lemma. *Let $k = \operatorname{rank}(M), v \in \mathcal{R}_k$ and $\xi = \gamma_v(-\infty)$, $\eta = \gamma_v(\infty)$. Then there exists an $\varepsilon > 0$ such that if F is a k-flat in M with $d(p, F) = 1$, where $p = \pi(v)$ is the foot point of v, then*

$$\angle(\xi, F(\infty)) + \angle(\eta, F(\infty)) \geq \varepsilon\,.$$

Proof. Suppose there is no $\varepsilon > 0$ with the asserted properties. Then, by the semi-continuity of $\angle$, there is a k-flat F in M with $d(p, F) = 1$ and $\xi, \eta \in F(\infty)$. Since F is a flat and $\angle(\xi, \eta) = \pi$ this implies $\angle_q(\xi, \eta) = \pi$ for all $q \in F$. Hence F consists of geodesics parallel to γ_v and therefore $F \subset P(\gamma_v) = P_v$. But P_v is a k-flat, hence $F = P_v$. This is a contradiction since $d(p, F) = 1$. $\square$

5.3 Lemma. *Let $k = \operatorname{rank} M$. Let $v \in \mathcal{R}_k$ be Γ-recurrent and let $\eta = \gamma_v(\infty)$. Then there exists an $\alpha > 0$ such that $\angle(\eta, \zeta) \geq \alpha$ for any $\zeta \in M(\infty) \setminus P_v(\infty)$.*

Proof. Let $\varepsilon > 0$ be as in Lemma 5.2. Choose $\delta > 0$ such that any unit vector w at the foot point p of v with $\angle(v, w) < \delta$ belongs to $\mathcal{R}_k$ and set $\alpha = \min\{\varepsilon, \delta\}$.

Let ζ be any point in $M(\infty) \setminus P_v(\infty)$ and suppose $\angle(\eta, \zeta) < \alpha$. In particular, $\angle(\eta, \zeta) < \pi$ and hence there is a unique $\angle$-geodesic $\sigma : [0,1] \to M(\infty)$ from η to ζ. Let $v(s)$ be the vector at p pointing at $\sigma(s), 0 \leq s \leq 1$. Note that

$$\angle(v(s), v(t)) \;\leq\; \angle(\sigma(s), \sigma(t)) \;=\; |t - s| \angle(\eta, \zeta)$$

by the definition of the $\angle$-metric on $M(\infty)$. Hence

$$\angle(v, v(s)) \le \angle(\eta, \sigma(s)) < \delta$$

and hence $v(s) \in \mathcal{R}_k$, $0 \le s \le 1$.

Let $w = v(1)$ be the vector at p pointing at ζ. Note that $\eta \notin P_w(\infty)$. Otherwise we would have $\gamma_v \subset P_w$ since P_w is convex and $p \in P_w$. But then $P_v = P_w$, a contradiction since $\zeta \notin P_v(\infty)$. We conclude that (the convex function) $d(\gamma_v(t), P_w) \to \infty$ as $t \to \infty$.

Since v is Γ-recurrent, there exist sequences $t_n \to \infty$ and (φ_n) in Γ such that $d\varphi_n(g^{t_n}v) \to v$ as $n \to \infty$. By what we just said, there is an $N \in \mathbb{N}$ such that $d(\gamma_v(t_n), P_w) \ge 1$ for all $n \ge N$. By Lemma 5.1 there exists an $s_n \in [0,1]$ with $d(\gamma_v(t_n), P_{v(s_n)}) = 1$, $n \ge N$. We let $F_n = \varphi_n(P(v_{s_n}))$. Then $d(p, F_n) \to 1$. Passing to a subsequence if necessary, the sequence (F_n) converges to a k-flat F with $d(p, F) = 1$.

Let γ_n be the geodesic defined by $\gamma_n(t) = \varphi_n(\gamma_v(t + t_n)), t \in \mathbb{R}$. Then $d(\gamma_n(t), F_n)$ is convex in t and increases from 0 to 1 on the interval $[-t_n, 0]$. By the choice of t_n and φ_n we have $\gamma_n \to \gamma_v$ and hence $d(\gamma_v(t), F) \le 1$ on $(-\infty, 0]$. We conclude that $\gamma_v(-\infty) \in F(\infty)$. Therefore $\angle(\gamma_v(\infty)), F(\infty)) \ge \varepsilon$, where $\varepsilon > 0$ is the constant from Lemma 5.2.

Let $\zeta_n = \sigma(s_n)$. By passing to a further subsequence if necessary, we can assume that $\varphi_n(\zeta_n) \in F_n(\infty)$ converges. Then the limit ξ is in $F(\infty)$. Also note that $\varphi_n(\eta) = \gamma_n(\infty)$ converges to $\eta = \gamma_v(\infty)$ since $\gamma_n \to \gamma_v$. Hence

$$\begin{aligned}\varepsilon &\le \angle(\gamma_v(\infty), F(\infty)) \le \angle(\gamma_v(\infty), \xi) \\ &\le \liminf \angle(\varphi_n(\eta), \varphi_n(\zeta_n)) \\ &= \liminf \angle(\eta, \zeta_n) \le \angle(\eta, \zeta) < \alpha.\end{aligned}$$

This is a contradiction. □

For any $p \in M$, let $\varphi_p : M(\infty) \to S_pM$ be the homeomorphism defined by $\varphi_p(\xi) = v_\xi(p)$, where $v_\xi(p)$ is the unit vector at p pointing at ξ.

5.4 Lemma. *Let v, η and α be as in Lemma 5.3, and let F be a k-flat with $\eta = \gamma_v(\infty) \in F(\infty)$. Then*

$$B_\alpha(\eta) = \{\zeta \in M(\infty) \mid \angle(\eta, \zeta) < \alpha\} \subset F(\infty)$$

and for any $q \in F$, φ_q maps $B_\alpha(\eta)$ isometrically onto the ball $B_\alpha(v_\eta(q))$ of radius α in S_qF.

Proof. Lemma 5.3 implies $B_\alpha(\eta) \subset P_v(\infty)$. For $q \in F$, φ_q^{-1} maps S_qF isometrically into $M(\infty)$. In particular, φ_q^{-1} maps $B_\alpha(v_\eta(q))$ isometrically into $B_\alpha(\eta)$. Hence $\varphi_p \circ \varphi_q^{-1} : B_\alpha(v_\eta(q)) \to B_\alpha(v)$ is an isometric embedding mapping $v_\eta(q)$ to v. Since both, $B_\alpha(v_\eta(q))$ and $B_\alpha(v)$, are balls of radius α in a unit sphere of dimension $k-1$, $\varphi_p \circ \varphi_q^{-1}$ is surjective and hence an isometry. The lemma follows. □

5.5 Corollary. *Let v, η and α be as in Lemma 5.3. Then $\angle_q(\zeta_1, \zeta_2) = \angle(\zeta_1, \zeta_2)$ for all $q \in M$ and $\zeta_1, \zeta_2 \in B_\alpha(\eta)$.*

Proof. Let $q \in M$ and let γ be the unit speed geodesic through q with $\gamma(\infty) = \eta$. By Proposition 4.4, γ is contained in a k-flat F. By Lemma 5.4, we have $B_\alpha(\eta) \subset F(\infty)$. Since F is a flat we have $\angle_q(\zeta_1, \zeta_2) = \angle(\zeta_1, \zeta_2)$ for $\zeta_1, \zeta_2 \in B_\alpha(\eta)$. □

5.6 Proposition. *If* rank$(M) = k \geq 2$, *if the isometry group Γ of M satisfies the duality condition and if M is not flat, then $M(\infty)$ contains a proper, nonempty, closed and Γ-invariant subset.*

Proof. Suppose first that M has a non-trivial Euclidean de Rahm factor M_0. Then $M = M_0 \times M_1$ with M_1 non-trivial since M is not flat. But then every isometry of M leaves this splitting invariant, and hence $M_0(\infty)$ and $M_1(\infty)$ are subsets of $M(\infty)$ as asserted. Hence we can assume from now on that M has no Euclidean de Rahm factor.

Let X_δ be the set of $\xi \in M(\infty)$ such that there is an $\eta \in M(\infty)$ with $\angle_q(\xi, \eta) = \delta$ for all $q \in M$. Corollary 5.5 shows that $X_\delta \neq \emptyset$ for some $\delta > 0$. Clearly, X_δ is closed and Γ-invariant for any δ.

Let $\beta = \sup\{\delta \mid X_\delta \neq \emptyset\}$. Then $X_\beta \neq \emptyset$. We want to show $X_\beta \neq M(\infty)$. We argue by contradiction and assume $X_\beta = M(\infty)$. Suppose furthermore $\beta < \pi$. Let $v \in \mathcal{R}_k$ be Γ-recurrent, $\eta = \gamma_v(\infty)$ and let α be as in Lemma 5.3. Choose $\delta > 0$ with $\beta + \delta < \pi$ and $\delta < \alpha$. Since $X_\beta = M(\infty)$, there is a point $\zeta \in M(\infty)$ with $\angle_q(\eta, \zeta) = \beta < \pi$ for all $q \in M$. Let $\sigma : [0, \beta] \to M(\infty)$ be the unique unit speed $\angle$-geodesic from ζ to η. Now the last piece of σ is in $B_\alpha(\eta)$ and $B_\alpha(\eta)$ is isometric to a ball of radius α in a unit sphere. Hence σ can be prolongated to a $\angle$-geodesic $\sigma^* : [0, \beta + \delta] \to M(\infty)$.

Let $q \in M$. Since $\angle(\varphi_q(\xi_1), \varphi_q(\xi_2)) \leq \angle(\xi_1, \xi_2)$ for all $\xi_1, \xi_2 \in M(\infty)$, where $\varphi_q(\xi) \in S_q M$ denotes the unit vector at q pointing at $\xi \in M(\infty)$, and since $\angle_q(\eta, \zeta) = \beta = \angle(\eta, \zeta)$, we conclude

$$\angle(\varphi_q(\sigma(s)), \varphi_q(\sigma(t))) = \angle(\sigma(s), \sigma(t)) = |s - t|$$

for all $s, t \in [0, \beta]$. Hence $\varphi_q \circ \sigma$ is a unit speed geodesic in $S_q M$.

The geodesic γ through q with $\gamma(\infty) = \eta$ is contained in a k-flat F, and hence $\varphi_q \circ \sigma^* : (\beta - \alpha, \beta + \delta]$ is also a unit speed geodesic in $S_q M$. Therefore $\varphi_q \circ \sigma^*$ is a unit speed geodesic in $S_q M$ of length $\beta + \delta$ and hence

$$\angle_q(\zeta, \sigma^*(\beta + \delta)) = \beta + \delta\,.$$

Since q was chosen arbitrary we get $X_{\beta+\delta} \neq \emptyset$. This contradicts the choice of β, hence $\beta = \pi$. But $X_\pi \neq \emptyset$ implies that M is isometric to $M' \times \mathbb{R}$. This is a contradiction to the assumption that M does not have a Euclidean factor. Hence X_β is a proper, nonempty, closed and Γ-invariant set as claimed. □

5.7 Remark. The set X_β as in the proof catches endpoints of (certain) singular geodesics in the case that X is a symmetric space of non-compact type.

6. Proof of the rank rigidity

From now on we assume that M is a Hadamard manifold of rank ≥ 2 and that the isometry group of M satisfies the duality condition. In the last section we have seen that there is a Γ-invariant proper, nonempty compact subset $Z \subset M(\infty)$. By Theorem III.2.4 and Formula III.2.5.

$$(6.1) \qquad f(v) = \angle(v, Z) := \inf\{\angle_{\pi(v)}(\gamma_v(\infty), \xi) \mid \xi \in Z\}$$

defines a Γ-invariant continuous first integral for the geodesic flow on SM and

$$(6.2) \qquad f(v) = f(w) \quad \text{if} \quad \gamma_v(\infty) = \gamma_w(\infty) \quad \text{or} \quad \gamma_v(-\infty) = \gamma_w(-\infty).$$

6.3 Lemma. *$f : SM \to \mathbb{R}$ is locally Lipschitz, hence differentiable almost everywhere.*

Proof. Instead of the canonical Riemannian distance d_S on SM, we consider the metric

$$d_1(v, w) = d(\gamma_v(0), \gamma_w(0)) + d(\gamma_v(1), \gamma_w(1))$$

which is locally equivalent to d_S. Let $v, w \in SM$ and $p = \pi(v), q = \pi(w)$. Let v' be the vector at q asymptotic to v. Then $f(v') = f(v)$ by (6.2). Now

$$\begin{aligned} d_1(w, v') &= d(\gamma_w(1), \gamma_{v'}(1)) \leq d(\gamma_w(1), \gamma_v(1)) + d(\gamma_v(1), \gamma_{v'}(1)) \\ &\leq d(\gamma_w(1), \gamma_v(1)) + d(\gamma_v(0), \gamma_{v'}(0)) = d_1(v, w) \end{aligned}$$

since v and v' are asymptotic and $\gamma_{v'}(0) = \gamma_w(0)$. By comparison we obtain

$$\begin{aligned} |f(v) - f(w)| &= |f(v') - f(w)| \\ &\leq \angle(v', w) \leq 2 \arcsin\left(\frac{d_1(v, w)}{2}\right). \qquad \square \end{aligned}$$

Recall the definition of the sets $W^{so}(v)$, $v \in SM$, at the end of Section 3,

$$W^{so}(v) = \{w \in SM \mid w(\infty) = v(\infty)\}.$$

Correspondingly we define

$$W^{uo}(v) = \{w \in SM \mid w(-\infty) = v(-\infty)\}.$$

Note that $W^{uo}(v) = -W^{so}(-v)$. One of the central observations in [Ba2] is as follows.

6.4 Lemma. *Let $v \in SM$. Then $T_v W^{so}(v) + T_v W^{uo}(v)$ contain the horizontal subspace of $T_v SM$.*

Proof. Note that $T_v W^{so}(v)$ consists of all pairs of the form $(X, B^+(X))$, where $B^+(X)$ denotes the covariant derivative of the stable Jacobi field along γ_v determined by X. It follows that $T_v W^{uo}(v)$ consists of all pairs of the form $(X, B^-(X))$, where $B^-(X)$ denotes the covariant derivative of the unstable Jacobi field along γ_v determined by X. Observe that B^+ and B^- are symmetric linear operators of $T_{\pi(v)}M$, and

$$E_0 = \{X \in T_{\pi(v)}M \mid B^+(X) = B^-(X) = 0\}$$

consists of all X which determine a parallel Jacobi field along γ_v. (In particular, $v \in E_0$). Since B^+ and B^- are symmetric, they map $T_{\pi(v)}M$ into the orthogonal complement $E_0^\perp$ of E_0.

Given a horizontal vector $(X, 0)$, we want to write it in the form

$$(X, 0) = (X_1, B^+(X_1)) + (X_2, B^-(X_2))$$

which is equivalent to

$$X_2 = X - X_1 \text{ and } B^+(X_1) = -B^-(X_2).$$

Because of the fact that

$$B^-(X_2) = B^-(X) - B^-(X_1)$$

it suffices to solve

$$B^+(X_1) - B^-(X_1) = -B^-(X).$$

Since $B^-(X)$ is contained in $E_0^\perp$ and $(B^+ - B^-)(E_0^\perp) \subset E_0^\perp$, the latter equation has a solution if $(B^+ - B^-) \mid E_0^\perp$ is an isomorphism. This is clear, however, since $(B^+ - B^-)(Y) = 0$ implies that the stable and unstable Jacobi fields along γ_v determined by Y have identical covariant derivative in $\gamma_v(0)$; hence $Y \in E_0$. □

6.5 Corollary. *If f is differentiable at v and X is a horizontal vector in T_vSM, then $df_v(X) = 0$.*

Proof. By (6.2), f is constant on $W^{so}(v)$ and $W^{uo}(v)$. By Lemma 6.4 we have $X \in T_vW^{so}(v) + T_vW^{uo}(v)$. □

6.6 Lemma. *If c is a piecewise smooth horizontal curve in SM then $f \circ c$ is constant.*

Proof. Since f is continuous, it suffices to prove this for a dense set (in the C^0-topology) of piecewise smooth horizontal curves. Now f is differentiable in a set $D \subset SM$ of full measure. It is easy to see that we can approximate c locally by piecewise smooth horizontal curves $\tilde{c}$ such that $\tilde{c}(t) \in D$ for almost all t in the domain of $\tilde{c}$. Corollary 6.5 implies that $f \circ \tilde{c} = \text{const.}$ for such a curve $\tilde{c}$. □

Let $p \in M$. Since the subset $Z \subset M(\infty)$ is proper, the restriction of f to S_pM is not constant. Lemma 6.6 implies that f is invariant under the holonomy group G of M at p. In particular, the action of G on S_pM is not transitive. Now the theorem of Berger and Simons [Be, Si] applies and shows that M is a Riemannian product or a symmetric space of higher rank. This concludes the proof of the rank rigidity.

Appendix. Ergodicity of the geodesic flow

M. Brin

We present here a reasonably self-contained and short proof of the ergodicity of the geodesic flow on a compact n-dimensional manifold of strictly negative sectional curvature. The only complete proof of specifically this fact was published by D.Anosov in [Ano1]. The ergodicity of the geodesic flow is buried there among quite a number of general properties of hyperbolic systems, heavily depends on most of the rest of the results and is rather difficult to understand, especially for somebody whose background in smooth ergodic theory is not very strong. Other works on this issue either give only an outline of the argument or deal with a considerably more general situation and are much more difficult.

1. Introductory remarks

E.Hopf (see [Ho1], [Ho2]) proved the ergodicity of the geodesic flow for compact surfaces of variable negative curvature and for compact manifolds of constant negative sectional curvature in any dimension. The general case was established by D.Anosov and Ya.Sinai (see [Ano1], [AnSi]) much later. Hopf's argument is relatively short and very geometrical. It is based on the property of the geodesic flow which Morse called "instability" and which is now commonly known as "hyperbolicity". Although the later proofs by Anosov and Sinai follow the main lines of Hopf's argument, they are considerably longer and more technical. The reason for the over 30 year gap between the special and general cases is, in short, that length is volume in dimension 1 but not in higher dimensions. More specifically, these ergodicity proofs extensively use the horosphere $H(v)$ as a function of the tangent vector v and the absolute continuity of a certain map p between two horospheres (the absolute continuity of the horospheric foliations). Since the horosphere $H(v)$ depends on the infinite future of the geodesic γ_v determined by v, the function $H(\cdot)$ is continuous but, in the general case, not differentiable even if the Riemannian metric is analytic. As a result, the map p is, in general, only continuous. For $n = 2$ the horospheres are 1-dimensional and the property of bounded volume distortion for p is equivalent to its Lipschitz continuity which holds true in special cases but is false in general.

The general scheme of the ergodicity argument below is the same as Hopf's. Most ideas in the proof of the absolute continuity of the stable and unstable foliations are the same as in [Ano1] and [AnSi]. However, the quarter century development of hyperbolic dynamical systems and a couple of original elements have made the argument less painful and much easier to understand. The proof goes through with essentially no changes for the case of a manifold of negative sectional curvature with finite volume, bounded and bounded away from 0 curvature and bounded first derivatives of the curvature. For surfaces this is exactly Hopf's result. The assumption of bounded first derivatives cannot be avoided in this argument.

The proof also works with no changes for any manifold whose geodesic flow is Anosov.

In addition to being ergodic, the geodesic flow on a compact manifold of negative sectional curvature is mixing of all orders, is a K-flow, is a Bernoulli flow and has a countable Lebesgue spectrum, but we do not address these stronger properties here.

In Section 2 we list several basic facts from measure and ergodic theory. Continuous foliations with C^1-leaves and their absolute continuity are discussed in Section 3. In Section 4 we recall the definition of an Anosov flow and some of its basic properties and prove the Hölder continuity of the stable and unstable distributions. The absolute continuity of the stable and unstable foliations for the geodesic flow and its ergodicity are proved in Section 5.

2. Measure and ergodic theory preliminaries

A *measure space* is a triple $(X, \mathfrak{A}, \mu)$, where X is a set, $\mathfrak{A}$ is a σ-algebra of measurable sets and μ is a σ-additive measure. We will always assume that the measure is finite or σ-finite. Let $(X, \mathfrak{A}, \mu)$ and $(Y, \mathfrak{B}, \nu)$ be measure spaces. A map $\psi : X \to Y$ is called *measurable* if the preimage of any measurable set is measurable; it is called *nonsingular* if, in addition, the preimage of a set of measure 0 has measure 0. A nonsingular map from a measure space into itself is called a *nonsingular transformation* (or simply a *transformation*). Denote by λ the Lebesgue measure on $\mathbb{R}$. A measurable flow ϕ in a measure space $(X, \mathfrak{A}, \mu)$ is a map $\phi : X \times \mathbb{R} \to X$ such that (i) ϕ is measurable with respect to the product measure $\mu \times \lambda$ on $X \times \mathbb{R}$ and measure μ on X and (ii) the maps $\phi^t(\cdot) = \phi(\cdot, t) : X \to X$ form a one-parameter group of transformations of X, that is $\phi^t \circ \phi^s = \phi^{t+s}$ for any $t, s \in \mathbb{R}$. The general discussion below is applicable to the discrete case $t \in \mathbb{Z}$ and to a measurable action of any locally compact group.

A measurable function $f : X \to \mathbb{R}$ is *ϕ-invariant* if $\mu(\{x \in X : f(\phi^t x) \neq f(x)\}) = 0$ for every $t \in \mathbb{R}$. A measurable set $B \in \mathfrak{A}$ is *ϕ-invariant* if its characteristic function $\mathbf{1}_B$ is ϕ-invariant. A measurable function $f : X \to \mathbb{R}$ is *strictly* ϕ-invariant if $f(\phi^t x) = f(x)$ for all $x \in X$, $t \in \mathbb{R}$. A measurable set B is *strictly* ϕ-invariant if $\mathbf{1}_B$ is strictly ϕ-invariant.

We say that something holds true *mod 0* in X or for *μ-a.e.* x if it holds on a subset of full μ-measure in X. A *null* set is a set of measure 0.

2.1 Proposition. *Let ϕ be a measurable flow in a measure space $(X, \mathfrak{A}, \mu)$ and let $f : X \to \mathbb{R}$ be a ϕ-invariant function.*

Then there is a strictly ϕ-invariant measurable function $\tilde{f}$ such that $f(x) = \tilde{f}(x)$ mod 0.

Proof. Consider the measurable map $\Phi : X \times \mathbb{R} \to X$, $\Phi(x,t) = \phi^t x$, and the product measure ν in $X \times \mathbb{R}$. Since f is ϕ-invariant, $\nu(\{(x,t) : f(\phi^t x) = f(x)\}) = 1$. By the Fubini theorem, the set

$$A_f = \{x \in X : f(\phi^t x) = f(x) \text{ for a.e. } t \in \mathbb{R}\}$$

has full μ-measure. Set

$$\tilde{f}(x) = \begin{cases} f(y) & \text{if } \phi^t x = y \in A_f \text{ for some } t \in \mathbb{R} \\ 0 & \text{otherwise}. \end{cases}$$

If $\phi^t x = y \in A_f$ and $\phi^s x = z \in A_f$ then clearly $f(y) = f(z)$. Therefore $\tilde{f}$ is well defined and strictly ϕ-invariant. $\square$

2.2 Definition (Ergodicity). A measurable flow ϕ in a measure space $(X, \mathfrak{A}, \mu)$ is called *ergodic* if any ϕ-invariant measurable function is constant mod 0, or equivalently, if any measurable ϕ-invariant set has either full or 0 measure.

The equivalence of the two definitions follows directly from the density of the linear combinations of step functions in bounded measurable functions.

A measurable flow ϕ is *measure preserving* (or μ is *ϕ-invariant*) if $\mu(\phi^t(B)) = \mu(B)$ for every $t \in \mathbb{R}$ and every $B \in \mathfrak{A}$.

2.3 Ergodic Theorems. *Let ϕ be a measure preserving flow in a finite measure space $(X, \mathfrak{A}, \mu)$. For a measurable function $f : X \to \mathbb{R}$ set*

$$f_T^+(x) = \frac{1}{T}\int_0^T f(\phi^t x)\,dt \quad \text{and} \quad f_T^-(x) = \frac{1}{T}\int_0^T f(\phi^{-t} x)\,dt.$$

(1) *(Birkhoff) If $f \in L^1(X, \mathfrak{A}, \mu)$ then the limits $f^+(x) = \lim_{T\to\infty} f_T^+(x)$ and $f^-(x) = \lim_{T\to\infty} f_T^-(x)$ exist and are equal for μ-a.e. $x \in X$, moreover f^+, f^- are μ-integrable and ϕ-invariant.*

(2) *(von Neumann) If $f \in L^2(X, \mathfrak{A}, \mu)$ then f_T^+ and f_T^- converge in $L^2(X, \mathfrak{A}, \mu)$ to ϕ-invariant functions f^+ and f^-, respectively.*

$\square$

2.4 Remark. If $f \in L^2(X, \mathfrak{A}, \mu)$, that is, if f^2 is integrable, then f^+ and f^- represent the same element $\bar{f}$ of $L^2(X, \mathfrak{A}, \mu)$ and $\bar{f}$ is the projection of f onto the subspace of ϕ-invariant L^2-functions. To see this let h be any ϕ-invariant L^2-function.

Since ϕ preserves μ we have

$$\begin{aligned}\int_X (f(x)-\bar{f}(x))h\,d\mu(x) &= \int_X f(x)h(x)\,d\mu(x) - \int_X \lim_{T\to\infty}\frac{1}{T}\int_0^T f(\phi^t x)h(x)\,dt\,d\mu(x)\\ &= \int_X f(x)h(x)\,d\mu(x) - \int_X \lim_{T\to\infty}\frac{1}{T}\int_0^T f(y)h(\phi^{-1}y)\,dt\,d\mu(y)\\ &= \int_X f(x)h(x)\,d\mu(x) - \int_X f(y)h(y)\,d\mu(y) = 0.\end{aligned}$$

In the topological setting invariant functions are mod 0 constant on sets of orbits converging forward or backward in time.

Let $(X,\mathfrak{A},\mu)$ be a finite measure space such that X is a compact metric space with distance d, μ is a measure positive on open sets and $\mathfrak{A}$ is the μ-completion of the Borel σ-algebra. Let ϕ be a continuous flow in X, that is the map $(x,t)\to\phi^t x$ is continuous. For $x\in X$ define the *stable* $V^s(x)$ and *unstable* $V^u(x)$ sets by the formulas

$$\begin{aligned}V^s(x) &= \{y\in X \,:\, d(\phi^t x,\phi^t y)\to 0 \text{ as } t\to\infty\}\,,\\ V^u(x) &= \{y\in X \,:\, d(\phi^t x,\phi^t y)\to 0 \text{ as } t\to-\infty\}\,.\end{aligned} \tag{2.5}$$

2.6 Proposition. *Let ϕ be a continuous flow in a compact metric space X preserving a finite measure μ positive on open sets and let $f: X\to\mathbb{R}$ be a ϕ-invariant measurable function.*

Then f is mod 0 constant on stable sets and unstable sets, that is, there are null sets N_s and N_u such that $f(y)=f(x)$ for any $x,y\in X\setminus N_s$, $y\in V^s(x)$ and $f(z)=f(x)$ for any $x,z\in X\setminus N_u$, $z\in V^u(x)$.

Proof. We will only deal with the stable sets. Reversing the time gives the same for the unstable sets. Assume WLOG that f is nonnegative.

For $C\in\mathbb{R}$ set $f_C(x)=\min(f(x),C)$. Clearly f_C is ϕ-invariant and it is sufficient to prove the statement for f_C with arbitrary C. For a natural m let $h_m: X\to\mathbb{R}$ be a continuous function such that $\int_X |f_C-h_m|\,d\mu(x)<\frac{1}{m}$. By the Birkhoff Ergodic Theorem,

$$h_m^+(x)=\lim_{T\to\infty}\frac{1}{T}\int_0^T f_m(\phi^t x)\,dt$$

exists a.e. Since μ and f_C are ϕ-invariant, we have that for any $t\in\mathbb{R}$

$$\begin{aligned}\frac{1}{m} > \int_X |f_C(x)-h_m(x)|\,d\mu(x) &= \int_X |f_C(\phi^t y)-h_m(\phi^t y)|\,d\mu(y)\\ &= \int_X |f_C(y)-h_m(\phi^t y)|\,d\mu(y)\,.\end{aligned}$$

Therefore

$$\int_X \left| f_C(y) - \frac{1}{T}\int_0^T h_m(\phi^t y) \right| dt\, d\mu(y) \le \frac{1}{T}\int_0^T \int_X |f_C(y) - h_m(\phi^t y)|\, d\mu(y)\, dt < \frac{1}{m}.$$

Note that since h_m is uniformly continuous, $h_m^+(y) = h_m^+(x)$ whenever $y \in V^s(x)$ and $h_m^+(x)$ is defined. Hence there is a set N_m of μ-measure 0 such that h_m^+ exists and is constant on the stable sets in $X \setminus N_m$. Therefore, $f_C^+(x) := \lim_{m\to\infty} h_m^+(x)$ is constant on the stable sets in $X \setminus (\cup N_m)$. Clearly $f_C(x) = f_C^+(x)$ mod 0. □

For differentiable hyperbolic flows in general and for the geodesic flow on a negatively curved manifold in particular, the stable and unstable sets are differentiable submanifolds of the ambient manifold M^n. The foliations W^s and W^u into stable and unstable manifolds are called the stable and unstable foliations, respectively. Together with the 1-dimensional foliation W^o into the orbits of the flow they form three transversal foliations W^s, W^u and W^o of total dimension n. Any mod 0 invariant function f is, by definition, mod 0 constant on the leaves of W^o and, by Proposition 2.6, is mod 0 constant on the leaves of W^s and W^u. To prove ergodicity one must show that f is mod 0 locally constant. This would follow immediately if one could apply a version of the Fubini theorem to the three foliations. Unfortunately the foliations W^s and W^u are in general not differentiable and not integrable (see the next section for definitions). The applicability of a Fubini-type argument depends on the absolute continuity of W^s and W^u which is discussed in the next section.

We will need the following simple lemma in the proof of Theorem 5.1.

2.7 Lemma. *Let $(X, \mathfrak{A}, \mu)$, $(Y, \mathfrak{B}, \nu)$ be two compact metric spaces with Borel σ-algebras and σ-additive Borel measures and let $p_n : X \to Y$, $n = 1, 2, \ldots$, $p : X \to Y$ be continuous maps such that*

(1) *each p_n and p are homeomorphisms onto their images,*
(2) *p_n converges to p uniformly as $n \to \infty$,*
(3) *there is a constant C such that $\nu(p_n(A)) \le C\mu(A)$ for any $A \in \mathfrak{A}$.*

Then $\nu(p(A)) \le C\mu(A)$ for any $A \in \mathfrak{A}$.

Proof. It is sufficient to prove the statement for an arbitrary open ball B in X. For $\delta > 0$ let B_δ denote the δ-interior of B. Then $p(B_\delta) \subset p_n(B)$ for n large enough, and hence, $\nu(p(B_\delta)) \le \nu(p_n(B)) \le C\mu(B)$. Observe now that $\nu(p(B_\delta)) \nearrow \nu(p(B))$ as $\delta \searrow 0$. □

3. Absolutely continuous foliations

Let M be a smooth n-dimensional manifold and let B^k denote the closed unit ball centered at 0 in $\mathbb{R}^k$. A partition W of M into connected k-dimensional C^1-submanifolds $W(x) \ni x$ is called a *k-dimensional C^0-foliation of M with C^1-leaves* (or simply a *foliation*) if for every $x \in M$ there is a neighborhood $U = U_x \ni x$ and a homeomorphism $w = w_x : B^k \times B^{n-k} \to U$ such that $w(0,0) = x$ and

(i) $w(B^k, z)$ is the connected component $W_U(w(0,z))$ of $W(w(0,z)) \cap U$ containing $w(0,z)$,

(ii) $w(\cdot, z)$ is a C^1-diffeomorphism of B^k onto $W_U(w(0,z))$ which depends continuously on $z \in B^{n-k}$ in the C^1-topology.

We say that W is a C^1-foliation if the homeomorphisms w_x are diffeomorphisms.

3.1 Exercise. Let W be a k-dimensional foliation of M and let L be an $(n-k)$-dimensional local transversal to W at $x \in M$, that is, $T_xM = T_xW(x) \oplus T_xL$. Prove that there is a neighborhood $U \ni x$ and a C^1 coordinate chart $w : B^k \times B^{n-k} \to U$ such that the connected component of $L \cap U$ containing x is $w(0, B^{n-k})$ and there are C^1-functions $f_y : B^k \to B^{n-k}$, $y \in B^{n-k}$ with the following properties:

(i) f_y depends continuously on y in the C^1-topology;

(ii) $w(\text{graph}\,(f_y)) = W_U(w(0,y))$.

We assume that M is a Riemannian manifold with the induced distance d. Denote by m the Riemannian volume in M and by m_N denote the induced Riemannian volume in a C^1-submanifold N.

3.2 Definition (Absolute continuity). Let L be a $(n-k)$-dimensional open (local) transversal for a foliation W, that is, $T_xL \oplus T_xW(x) = T_xM$ for every $x \in L$. Let $U \subset M$ be an open set which is a union of local "leaves", that is $U = \cup_{x\in L} W_U(x)$, where $W_U(x) \approx B^k$ is the connected component of $W(x) \cap U$ containing x.

The foliation W is called *absolutely continuous* if for any L and U as above there is a measurable family of positive measurable functions $\delta_x : W_U(x) \to \mathbb{R}$ (called the *conditional densities*) such that for any measurable subset $A \subset U$

$$m(A) = \int_L \int_{W_U(x)} \mathbf{1}_A(x,y)\delta_x(y)\, dm_{W(x)}(y)\, dm_L(x)\,.$$

Note that the conditional densities are automatically integrable. Instead of absolute continuity we will deal with "*transversal absolute continuity*" which is a stronger property, see Proposition 3.4.

3.3 Definition (Transversal absolute continuity). Let W be a foliation of M, $x_1 \in M$, $x_2 \in W(x_1)$ and let $L_i \ni x_i$ be two transversals to W, $i = 1,2$. There are neighborhoods $U_i \ni x_i$, $U_i \subset L_i$, $i = 1,2$, and a homeomorphism $p : U_1 \to U_2$ (called the *Poincaré map*) such that $p(x_1) = x_2$ and $p(y) \in W(y)$, $y \in U_1$. The foliation W is called *transversally absolutely continuous* if the Poincaré map p is absolutely continuous for any transversals L_i as above, that is, there is a positive measurable function $q : U_1 \to \mathbb{R}$ (called the *Jacobian* of p) such that for any measurable subset $A \subset U_1$

$$m_{L_2}(p(A)) = \int_{U_1} \mathbf{1}_A q(y) dm_{L_1}(y)\,.$$

If the Jacobian q is bounded on compact subsets of U_1 then W is called transversally absolutely continuous with bounded Jacobians.

3.4 Exercise. Is the Poincaré map in Definition 3.3 uniquely defined?.

3.5 Proposition. *If W is transversally absolutely continuous, then it is absolutely continuous.*

Proof. Let L and U be as in Definition 3.2, $x \in L$ and let F be an $(n-k)$-dimensional C^1-foliation such that $F(x) \supset L$, $F_U(x) = L$ and $U = \cup_{y \in W_U(x)} F_U(y)$, see Figure 2. Obviously F is absolutely continuous and transversally absolutely continuous. Let $\bar{\delta}_y(\cdot)$ denote the conditional densities for F. Since F is a C^1-foliation, $\bar{\delta}$ is continuous, and hence, measurable. For any measurable set $A \subset U$, by the Fubini theorem,

$$m(A) = \int_{W_U(x)} \int_{F_U(y)} \mathbf{1}_A(y,z) \bar{\delta}_y(z) \, dm_{F(y)}(z) \, dm_{W(x)}(y) \,. \tag{3.6}$$

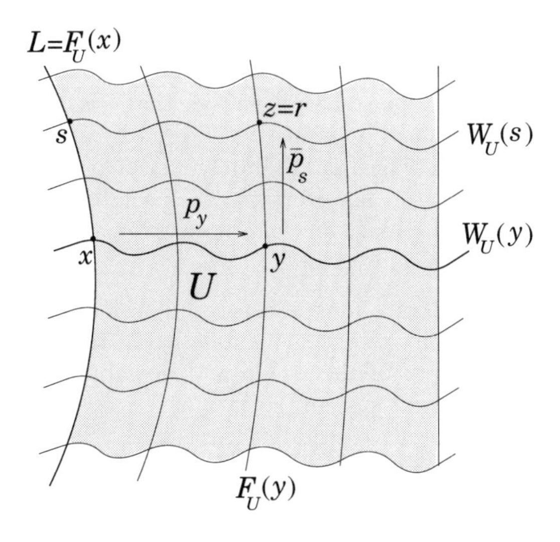

FIGURE 2

Let p_y denote the Poincaré map along the leaves of W from $F_U(x) = L$ to $F_U(y)$ and let $q_y(\cdot)$ denote the Jacobian of p_y. It is not difficult to see that q is a measurable function (see the exercise below). We have

$$\int_{F_U(y)} \mathbf{1}_A(y,z) \bar{\delta}_y(z) \, dm_{F(y)}(z) = \int_L \mathbf{1}_A(p_y(s)) q_y(s) \bar{\delta}_y(p_y(s)) \, dm_L(s)$$

and by changing the order of integration in (3.6) we get

$$m(A) = \int_L \int_{W_U(x)} \mathbf{1}_A(p_y(s)) q_y(s) \bar{\delta}_y(p_y(s))\, dm_{W(x)}(y)\, dm_L(s)\,. \tag{3.7}$$

Similarly, let $\bar{p}_s$ denote the Poincaré map along the leaves of F from $W_U(x)$ to $W_U(s)$, $s \in L$, and let $\bar{q}_s$ denote the Jacobian of $\bar{p}_s$. We transform the integral over $W_U(x)$ into an integral over $W_U(s)$ using the change of variables $r = p_y(s)$, $y = \bar{p}_s^{-1}(r)$

$$\int_{W_U(x)} \mathbf{1}_A(p_y(s)) q_y(s) \bar{\delta}_y(p_y(s)) dm_{W(x)}(y) = \int_{W_U(s)} \mathbf{1}_A(r) q_y(s) \bar{\delta}_y(r) \bar{q}_s^{-1}(r) dm_{W(s)}(r).$$

The last formula together with (3.7) gives the absolute continuity of W. □

3.8 Exercise. Prove that the Jacobian q in the above argument is a measurable function.

3.9 Remark. The converse of Proposition 3.5 is not true in general. To see this imagine two parallel vertical intervals I_1 and I_2 in the plane making the opposite sides of the unit square. Let C_1 be a "thick" Cantor set in I_1, that is, a Cantor set of positive measure, and let C_2 be the standard "1/3" Cantor set in I_2. Let $\alpha : I_1 \to I_2$ be an increasing homeomorphism such that α is differentiable on $I_1 \setminus C_1$ and $\alpha(C_1) = C_2$. Connect each point $x \in I_1$ to $\alpha(x)$ by a straight line to obtain a foliation W of the square which is absolutely continuous but not transversally absolutely continuous. It is true, however, that an absolutely continuous foliation is transversally absolutely continuous for "almost every" pair of transversals chosen, say, from a smooth nondegenerate family.

3.10 Exercise. Fill in the details for the construction of an absolutely continuous but not transversally absolutely continuous foliation in the previous remark.

The absolute continuity of the stable and unstable foliations for a differentiable hyperbolic dynamical system, such as the geodesic flow on a compact Riemannian manifold of negative sectional curvature, is precisely the statement that allows one to use a Fubini-type argument and to conclude that since an invariant function is mod 0 constant on stable and unstable manifolds, it must be mod 0 constant in the phase space.

3.11 Lemma. *Let W be an absolutely continuous foliation of a Riemannian manifold M and let $f : M \to \mathbb{R}$ be a measurable function which is mod 0 constant on the leaves of W.*

Then for any transversal L to W there is a measurable subset $\widetilde{L} \subset L$ of full induced Riemannian volume in L such that for every $x \in \widetilde{L}$ there is a subset $\widetilde{W}(x) \subset W(x)$ of full induced Riemannian volume in $W(x)$ on which f is constant.

Proof. Given a transversal L break it into smaller pieces L_i and consider neighborhoods U_i as in the definition of absolute continuity. Let N be the null set such that f is constant on the leaves of W in $M \setminus N$ and let $N_i = N \cap U_i$. By the absolute continuity of W for each i there is a subset $\widetilde{L}_i$ of full measure in L_i such that for each $x \in \widetilde{L}_i$ the complement $\widetilde{W}_{U_i}(x)$ of N_i in $W_{U_i}(x)$ has full measure in the local leaf. □

Two foliations W_1 and W_2 of a Riemannian manifold M are *transversal* if $T_xW_1(x) \cap T_xW_2(x) = \{0\}$ for every $x \in M$.

3.12 Proposition. *Let M be a connected Riemannian manifold and let W_1, W_2 be two transversal absolutely continuous foliations on M of complementary dimensions, that is $T_xM = T_xW_1(x) \oplus T_xW_2(x)$ for every $x \in M$ (the sum is direct but not necessarily orthogonal). Assume that $f : M \to \mathbb{R}$ is a measurable function which is mod 0 constant on the leaves of W_1 and mod 0 constant on the leaves of W_2.*

Then f is mod 0 constant in M.

Proof. Let $N_i \subset M$, $i = 1, 2$, be the null sets such that f is constant on the leaves of W_i in $M_i = M \setminus N_i$. Let $x \in M$ and let $U \ni x$ be a small neighborhood. By Lemma 3.11, since W_1 is absolutely continuous, there is y arbitrarily close to x whose local leaf $W_{U_1}(y)$ intersects M_1 by a set $\widetilde{M_1}(y)$ of full measure. Since W_2 is absolutely continuous, for almost every $z \in \widetilde{M_1}(y)$ the intersection $W_2 \cap M_2$ has full measure. Therefore f is mod 0 constant in a neighborhood of x. Since M is connected, f is mod 0 constant in M. □

The previous proposition can be applied directly to the stable and unstable foliations of a volume preserving Anosov diffeomorphism to prove its ergodicity (after establishing the absolute continuity of the foliations). The case of flows is more difficult since there are three foliations involved – stable, unstable and the foliation into the orbits of the flow. In that case one gets a pair of transversal absolutely continuous foliations of complementary dimensions by considering the stable foliation W^s and the weak unstable foliation W^{uo} whose leaves are the orbits of the unstable leaves under the flow.

Two transversal foliations W_i of dimensions d_i, $i = 1, 2$, are *integrable* and their *integral hull* is a $(d_1 + d_2)$-dimensional foliation W if

$$W(x) = \cup_{y \in W_1(x)} W_2(y) = \cup_{z \in W_2(x)} W_1(z) .$$

3.13 Lemma. *Let W_i, $i = 1, 2$, be transversal integrable foliations of a Riemannian manifold M with integral hull W such that W_1 is a C^1-foliation and W_2 is absolutely continuous.*

Then W is absolutely continuous.

Proof. Let L be a transversal for W. For a properly chosen neighborhood U in M the set $\widetilde{L} = \cup_{x \in L} W_{1U}(x)$ is a transversal for W_2. Since W_2 is absolutely continuous, we have for any measurable subset $A \subset U$

$$m(A) = \int_{\widetilde{L}} \int_{W_{2U}(y)} \mathbf{1}_A(y, z) \delta_y(z) \, dm_{W_{2U}(y)}(z) \, dm_{\widetilde{L}}(y) .$$

Since $\widetilde{L}$ is foliated by W_1-leaves,

$$\int_{\widetilde{L}} dm_{\widetilde{L}}(x) = \int_L \int_{W_{1U}(x)} j(x,y)\, dm_{1U(x)}(y)\, dm_L(x)$$

for some positive measurable function j. The absolute continuity of W follows immediately. □

We will need several auxiliary statements to prove the absolute continuity of the stable and unstable foliations of the geodesic flow.

4. Anosov flows and the Hölder continuity of invariant distributions

For a differentiable flow denote by W^o the foliation into the orbits of the flow and by E^o the line field tangent to W^o.

4.1 Definition (Anosov flow). A differentiable flow ϕ^t in a compact Riemannian manifold M is called *Anosov* if it has no fixed points and there are distributions E^s, $E^u \subset TM$ and constants $C, \lambda > 0$, $\lambda < 1$, such that for every $x \in M$ and every $t \geq 0$

$$(a)\ E^s(x) \oplus E^u(x) \oplus E^o(x) = T_xM\,,$$

$$(b)\ \|d\phi_x^t v_s\| \leq C\lambda^t \|v_s\| \quad \text{for any} \quad v_s \in E^s(x)\,,$$

$$(c)\ \|d\phi_x^{-t} v_u\| \leq C\lambda^t \|v_u\| \quad \text{for any} \quad v_u \in E^u(x)$$

It follows automatically from the definition that the distributions E^s and E^u (called the *stable* and *unstable* distributions, respectively) are continuous, invariant under the derivative $d\phi^t$ and their dimensions are (locally) constant and positive. To see this observe that if a sequence of tangent vectors $v_k \in TM$ satisfies (b) and $v_k \to v \in TM$ as $k \to \infty$, then v satisfies (b), and similarly for (c).

The stable and unstable distributions of an Anosov flow are integrable in the sense that there are stable and unstable foliations W^s and W^u whose tangent distributions are precisely E^s and E^u. The manifolds $W^s(x)$ and $W^u(x)$ are the stable and unstable sets as described in (2.5). This is proved in any book on differentiable hyperbolic dynamics. The geodesic flow on a manifold M of negative sectional curvature is Anosov. Its stable and unstable subspaces are the spaces of stable and unstable Jacobi fields perpendicular to the geodesic and the (strong) stable and unstable manifolds are unit normal bundles to the horospheres (see Chapter IV). If the sectional curvature of M is pinched between $-a^2$ and $-b^2$ with $0 < a < b$ then Inequalities (b) and (c) of Definition 4.1 hold true with $\lambda = e^{-a}$ (see Proposition IV.2.9).

For subspaces H_1, $H_2 \subset T_xM$ define the distance $\operatorname{dist}(H_1, H_2)$ as the Hausdorff distance between the unit spheres in H_1 and H_2. We say that H_2 is θ-*transversal* to H_1 if $\min \|v_1 - v_2\| \geq \theta$, where the minimum is taken over all unit vectors $v_1 \in H_1$, $v_2 \in H_2$.

Set $E^{so}(x) = E^s(x) \oplus E^o(x)$, $E^{uo}(x) = E^u(x) \oplus E^o(x)$. By compactness, the pairs $E^s(x)$, $E^{uo}(x)$ and $E^u(x)$, $E^{so}(x)$ are θ-transversal with some $\theta > 0$ independent of x.

4.2 Lemma. *Let ϕ^t be an Anosov flow. Then for every $\theta > 0$ there is $C_1 > 0$ such that for any subspace $H \subset T_xM$ with the same dimension as $E^s(x)$ and θ-transverssal to $E^{u0}(x)$ and any $t \geq 0$*

$$dist\,(d\phi^{-t}(x)H, E^s(\phi^{-t}x)) \leq C_1\lambda^t\, dist\,(H, E^s(x))\,.$$

Proof. Let $v \in H$, $\|v\| = 1$, $v = v_s + v_{uo}$, $v_s \in E^s(x)$, $v_{uo} \in E^{uo}(x)$. Then $\|v_s\| >$ const $\cdot\, \theta$, $d\phi^{-t}(x)v_s \in E^s(\phi^{-t}x)$, $d\phi^{-t}(x)v_{uo} \in E^{uo}(\phi^{-t}x)$ and $\|d\phi^{-t}(x)v_{uo}\| \leq$ const $\cdot\, \|v_{uo}\|$, $\|d\phi^{-t}(x)v_s\| \geq C^{-1}\lambda^{-t}\|v_s\|$. □

A distribution $E \subset TM$ is called *Hölder continuous* if there are constants $A, \alpha > 0$ such that for any $x, y \in M$

$$\mathrm{dist}(E(x), E(y)) \leq Ad(x,y)^\alpha\,,$$

where $\mathrm{dist}(E(x), E(y))$ denotes, for example, the Hausdorff distance in TM between the unit spheres in $E(x)$ and $E(y)$. The numbers A and α are called the Hölder constant and exponent, respectively.

4.3 Adjusted metric. Many arguments in hyperbolic dynamical systems become technically easier if one uses the so-called *adjusted* metric in M which is equivalent to the original Riemannian metric. Let $\phi^t : M \to M$ be an Anosov flow. Let $\beta \in (\lambda, 1)$. For $v_o \in E^o$, $v_s \in E^s$, $v_u \in E^u$ and $T > 0$ set

$$|v_o| = \|v_o\|\,;\; |v_s| = \int_0^T \frac{\|d\phi^\tau v_s\|}{\beta^\tau}\,d\tau,\; |v_u| = \int_0^T \frac{\|d\phi^{-\tau} v_s\|}{\beta^\tau}\,d\tau,$$

$$|v_o + v_s + v_u|^2 = |v_o|^2 + |v_s|^2 + |v_u|^2\,.$$

Note that since $\beta > \lambda$, the integrals $\int_0^\infty \frac{\|d\phi^\tau v_s\|}{\beta^\tau}\,d\tau$ and $\int_0^\infty \frac{\|d\phi^{-\tau} v_s\|}{\beta^\tau}\,d\tau$ converge. For $t > 0$ we have

$$\begin{aligned}|d\phi^t v_s| &= \int_0^T \frac{\|d\phi^{t+\tau} v_s\|}{\beta^\tau}\,d\tau = \beta^t \int_t^{T+t} \frac{\|d\phi^\tau v_s\|}{\beta^\tau}\,d\tau \\ &= \beta^t\left(|v_s| - \int_0^t \frac{\|d\phi^\tau v_s\|}{\beta^\tau}\,d\tau + \int_T^{T+t} \frac{\|d\phi^\tau v_s\|}{\beta^\tau}\,d\tau\right).\end{aligned}$$

For T large enough the second integral in parentheses is less than the first one and we get

$$|d\phi^t v_s| \leq \beta^t |v_s|$$

and similarly for v_u. Note that E^s, E^u and E^o are orthogonal in the adjusted metric.

The metric $|\cdot|$ is clearly smooth and equivalent to $\|\cdot\|$ and in the hyperbolicity inequalities of Definition 4.1 λ is replaced by β and C by 1.

In the next proposition we prove the Hölder continuity of E^s and E^u. This was first proved by Anosov (see [Ano2]). To make the exposition complete we present a similar but shorter argument here. It uses the main idea of a more general argument from [BrK1] (see Theorem 5.2). To eliminate a couple of constants we assume that the flow is C^2 but the argument works equally well for a C^1-flow whose derivative is Hölder continuous.

4.4 Proposition. *Distributions E^s, E^u, E^{so}, E^{uo} of a C^2 Anosov flow ϕ^t are Hölder continuous.*

Proof. Since E^o is smooth and since the direct sum of two transversal Hölder continuous distribution is clearly Hölder continuous, it is sufficient to prove that E^s and E^u are Hölder continuous. We consider only E^s, the Hölder continuity of E^u can be obtained by reversing the time. As we mentioned above, the hyperbolicity conditions (see Definition 4.1) are closed, and hence, all four distributions are continuous. Obviously a distribution which is Hölder continuous in some metric is Hölder continuous in any equivalent metric. Fix $\beta \in (\lambda, 1)$ and use the adjusted metric for β with an appropriate T. We will not attempt to get the best possible estimates for the Hölder exponent and constant here.

The main idea is to consider two very close points $x, y \in M$ and their images $\phi^m x$ and $\phi^m y$. If the images are reasonably close, the subspaces $E^s(\phi^m x)$ and $E^s(\phi^m y)$ are (finitely) close by continuity. Now start moving the subspaces "back" to x and y by the derivatives $d\phi^{-1}$. By the invariance of E^s under the derivative, $d\phi^{-k}E^s(\phi^m x) = E^s(\phi^{m-k}x)$. By Lemma 4.2, the image of $E^s(\phi^m y)$ under $d_{\phi^m x}\phi^{-k}$ is exponentially in k close to $E^s(\phi^{m-k}x)$. An exponential estimate on the distance between the corresponding images of x and y allows us to replace $d\phi^{-1}$ along the orbit of y by $d\phi^{-1}$ along the orbit of x.

To abbreviate the notation we use ϕ instead of ϕ^1. Fix $\gamma \in (0, 1)$, let $x, y \in M$ and choose q such that $\gamma^{q+1} < d(x, y) \le \gamma^q$. Let $D \ge \max \|d\phi\|$ and fix $\varepsilon > 0$. Let m be the integer part of $(\log \varepsilon - q \log \gamma)/\log D$. Then

$$d(\phi^i x, \phi^i y) \le d(x, y)D^i \le \gamma^q D^i \quad \text{and} \quad d(\phi^i x, \phi^i y) \le \varepsilon, \; i = 0, 1, ..., m.$$

Assume that γ is small enough so that m is large. Consider a system of small enough coordinate neighborhoods $U_i \supset V_i \ni \phi^i x$ identified with small balls in $T_{\phi^i x}M$ by diffeomorphisms with uniformly bounded derivatives and such that $\phi^{-1}V_i \subset U_{i-1}$. Assume that ε is small enough, so that $\phi^i y \in V_i$ for $i \le m$. We identify the tangent spaces $T_z M$ for $z \in V_i$ with $T_{\phi^i x}M$ and the derivatives $(d_z\phi)^{-1}$ with matrices which may act on tangent vectors with footpoints anywhere in V_i. In the argument below we estimate the distance between $E^s(x)$ and the parallel translate of $E^s(y)$ from y to x. Since the distance function between points is Lipschitz continuous, our estimate implies the Hölder continuity of E^s.

Let $v_y \in E^s(\phi^m y)$, $|v_y| > 0$, $v_k = d\phi^{-k}v_y = v_k^s + v_k^{uo}$, $v_s \in E^s(\phi^{m-k}x)$, $v_{uo} \in E^{uo}(\phi^{m-k}x)$, $k = 0, 1, ..., m,$. Fix $\kappa \in (\sqrt{\beta}, 1)$. We use induction on k to show that if ε is small enough then $\|v_k^{uo}\|/\|v_k^s\| \le \delta\kappa^k$ for $k = 0, ...m$ and a small $\delta > 0$.

For $k = 0$ the above inequality is satisfied for a small enough ε since E^s is continuous. Assume that the inequality holds true for some k. Set $(d_{\phi^{m-k}x}\phi)^{-1} = A_k$, $(d_{\phi^{m-k}y}\phi)^{-1} = B_k$. By the choice of D we have $\|A_k - B_k\| \leq \text{const}\cdot\gamma^q D^{m-k} =: \eta_k \leq \text{const}\cdot\varepsilon$, where the constant depends on the maximum of the second derivatives of ϕ. We have

$$|A_k v_k^s| \geq \beta^{-1}|v_k^s|,\ |A_k v_k^{uo}| \leq |v_k^{uo}|,$$

$$v_{k+1} = B_k v_k = A_k v_k + (B_k - A_k)v_k = A_k(v_k^s + v_k^{uo}) + (B_k - A_k)v_k\,.$$

Therefore

$$\frac{|v_{k+1}^{uo}|}{|v_{k+1}^s|} \leq \frac{|A_k v_k^{uo}| + \|A_k - B_k\|\cdot|v_k|}{|A_k v_k^s| - |(A_k - B_k)v_k|} \leq (\delta\kappa^k + \eta_k)\frac{|v_k|}{\beta^{-1}(|v_k| - |v_k^{uo}|) - \eta_k|v_k|}$$
$$\leq \left((\delta\kappa^k + \eta_k)\sqrt{\beta}\right)\left(\frac{|v_k|}{|v_k|(1 - \beta\eta_k - \delta\kappa^k)}\sqrt{\beta}\right).$$

By the choice of m and D above, $\eta_k = \text{const}\cdot\gamma^q D^{m-k} \leq \text{const}\cdot\varepsilon D^{-k}$, and hence the inequality for $k+1$ holds true provided D is big enough, ε is small enough, $\sqrt{\beta} < \kappa$ and δ is chosen so that the last factor is less than 1.

For $k = m$ we get that $|v_m^{uo}|/|v_m^s| \leq \delta\kappa^m$ and, by the choice of m,

$$\frac{|v_m^{uo}|}{|v_m^s|} \leq \text{const}\cdot\gamma^{-(q+1)\frac{\log\kappa}{\log D}} \leq \text{const}\cdot\text{dist}\,(x,y)^{-\frac{\log\kappa}{\log D}}\,.$$

Note that by varying v_y we obtain all vectors $v_m \in E^s(y)$. □

5. Proof of absolute continuity and ergodicity

We assume now that M is a compact boundaryless m-dimensional Riemannian manifold of strictly negative sectional curvature. For $v \in SM$ set $E^s(v) = \{(X,Y) : Y = J_X'(0)\}$, where J_X is the stable field along γ_v which is perpendicular to $\dot\gamma_v$ and such that $J_X(0) = X$. Set $E^u(v) = -E^s(-v)$. By Section IV.2, the distributions E^s and E^u are the stable and unstable distributions of the geodesic flow g^t in the sense of Definition 4.1. By Section IV.3, the distributions E^s and E^u are integrable and the leaves of the corresponding foliations W^s and W^u of SM are the two components of the unit normal bundles to the horospheres.

5.1 Theorem. *Let M be a compact Riemannian manifold with a C^3-metric of negative sectional curvature. Then the foliations W^s and W^u of SM into the normal bundles to the horospheres are transversally absolutely continuous with bounded Jacobians.*

Proof. We will only deal with the stable foliation W^s. Reverse the time for the geodesic flow g^t to get the statement for W^u. Let L_i be two C^1-transversals to W^s and let U_i and p be as in Definition 3.3. Denote by Σ_n the foliation of SM

into "inward" spheres, that is, each leaf of Σ_n is the set of unit vectors normal to a sphere of radius n in M and pointing inside the sphere. Let $V_i \subset U_i$ be closed subsets such that V_2 contains a neighborhood of $p(V_1)$. Denote by p_n the Poincaré map for Σ_n and transversals L_i restricted to V_1. We will use Lemma 16 to prove that the Jacobian q of p is bounded. By the construction of the horospheres, $p_n \rightrightarrows p$ as $n \to \infty$. We must show that the Jacobians q_n of p_n are uniformly bounded.

To show that the Jacobians q_n of the Poincaré maps $p_n : L_1 \to L_2$ are uniformly bounded we represent p_n as the following composition

$$p_n = g^{-n} \circ P_0 \circ g^n \,, \tag{5.2}$$

where g^n is the geodesic flow restricted to the transversal L_1 and $P_0 : g^n(L_1) \to g^n(L_2)$ is the Poincaré map along the vertical fibers of the natural projection $\pi : SM \to M$. For a large enough n, the spheres Σ_n are close enough to the stable horospheres W^s, and hence, are uniformly transverse to L_1 and L_2 so that p_n is well defined on V_1. Let $v_i \in V_i$, $i = 1, 2$, and $p_n(v_1) = v_2$. Then $\pi(g^n v_1) = \pi(g^n v_2)$ and $p_0(g^n v_1) = g^n v_2$. Let J_k^i denote the Jacobian of the time 1 map g^1 in the direction of $T_k^i = T_{g^k v_i} L_i$ at $g^k v_i \in L_i$, $i = 1, 2$, that is $J_k^i = |\det(dg^1(g^k v_i)|_{T_k^i})|$. If J_0 denotes the Jacobian of P_0, we get from (5.2)

$$q_n(v_1) = \prod_{k=0}^{n-1} (J_k^2)^{-1} \cdot J_0 \cdot \prod_{k=0}^{n-1} (J_k^1) = J_0(g^n v_1) \cdot \prod_{k=0}^{n-1} (J_k^1 / J_k^2) \,. \tag{5.3}$$

By Lemma 4.2, for a large n the tangent plane at $g^n v_1$ to the image $g^n(L_1)$ is close to $E^{so}(g^n v_1)$, and hence, is uniformly (in v_1) transverse to the unit sphere $S_x M$ at the point $x = \pi(g^n v_1) = \pi(g^n v_2)$. Note also that the unit spheres form a fixed smooth foliation of SM. Therefore the Jacobian J_0 is bounded from above uniformly in v_1.

We will now estimate from above the last product in (5.3). By Proposition IV.2.10,

$$d(g^k v_1, g^k v_2) \le \text{const} \cdot \exp(-ak) \cdot d(v_1, v_2) \,.$$

Hence, by the Hölder continuity of E^{uo},

$$\text{dist}(E^{uo}(g^k v_1), E^{uo}) \le \text{const} \cdot d(g^k v_1, g^k v_2)^\alpha \le \text{const} \cdot \exp(-ak\alpha) \,.$$

Together with Lemma 4.2 this implies that

$$\text{dist}(T_k^1, T_k^2) \le \text{const} \cdot \exp(-\beta k)$$

with some $\beta > 0$. Our assumptions on the metric imply that the derivative $dg^1(v)$ is Lipschitz continuous in v. Therefore its determinants in two exponentially close directions are exponentially close, that is $\|J_k^1 - J_k^2\| \leq \text{const} \cdot \exp(-\gamma k)$. Observe now that $\|J_k^i\|$ is uniformly separated away from 0 by the compactness of M. Therefore the last product in (5.3) is uniformly bounded in n and v_1. By Lemma 2.7, the Jacobian q of the Poincaré map p is bounded. □

We will need the following property of absolutely continuous foliations in the proof of Theorem 5.5.

5.4 Lemma. *Let W be an absolutely continuous foliation of a manifold Q and let $N \subset Q$ be a null set. Then there is a null set N_1 such that for any $x \in Q \setminus N_1$ the intersection $W(x) \cap N$ has conditional measure 0 in $W(x)$.*

Proof. Consider any local transversal L for W. Since W is absolutely continuous, there is a subset $\widetilde{L}$ of full m_L measure in L such that $m_{W(x)}(N) = 0$ for $x \in \widetilde{L}$. Clearly the set $\cup_{x \in \widetilde{L}} W(x)$ has full measure. □

5.5 Theorem. *Let M be a compact Riemannian manifold with a C^3 metric of negative sectional curvature. Then the geodesic flow $g^t : SM \to SM$ is ergodic.*

Proof. Let m be the Riemannian volume in M, λ_x be the Lebesgue measure in S_xM, $x \in M$ and μ be the Liouville measure in SM, $d\mu(x,v) = dm(x) \times d\lambda_x(v)$. Since M is compact, $m(M) < \infty$ and $\mu(SM) = m(M) \cdot \lambda_x(S_xM)$. The local differential equation determining the geodesic flow g^t is

$$\dot{x} = v, \quad \dot{v} = 0\,.$$

By the divergence theorem, the Liouville measure is invariant under g^t (see also Section IV.1). Hence the Birkhoff Ergodic Theorem and Proposition 2.6 hold true for g^t.

Since the unstable foliation W^u is absolutely continuous and the foliation W^o into the orbits of g^t is C^1, the integral hull W^{uo} is absolutely continuous by Lemma 3.13. Let $f : SM \to \mathbb{R}$ be a measurable g-invariant function. By Proposition 2.1, f can be corrected on a set of Liouville measure 0 to a strictly g-invariant function $\tilde{f}$. By Proposition 2.6, there is a null set N_u such that $\tilde{f}$ is constant on the leaves of W^u in $SM \setminus N_u$. Applying Lemma 5.4 twice we obtain a null set N_1 such that N_u is a null set in $W^{uo}(v)$ and in $W^u(v)$ for any $v \in SM \setminus N_1$. It follows that f is mod 0 constant on the leaf $W^u(v)$ and, since it is strictly g-invariant, is mod 0 constant on the leaf $W^{uo}(v)$. Hence f is mod 0 constant on the leaves of W^{uo}. Now apply Proposition 3.12 to W^s and W^{uo}. □

Bibliography

[AbS1] U. Abresch & V. Schroeder, *Graph manifolds, ends of negatively curved spaces and the hyperbolic* 120-*cell space*, J. Differential Geometry 35 (1992), 299–336.

[AbS2] U. Abresch & V. Schroeder, *Analytic manifolds of nonpositive curvature*, Preprint.

[ACBu] N. A'Campo & M. Burger, *Réseaux arithmétiques et commensurateur d'après G.A. Margulis*, Inventiones math. 116 (1994), 1–25.

[Ad] S. Adams, *Rank rigidity for foliations by manifolds of nonpositive curvature.*

[Alb1] S.I. Al'ber, *On n-dimensional problems in the calculus of variations in the large*, Soviet Math. Doklady 5 (1964), 700–804.

[Alb2] S.I. Al'ber, *Spaces of mappings into a manifold with negative curvature*, Soviet Math. Doklady 9 (1967), 6–9.

[Alek] D. V. Alekseevskiĭ, *Homogeneous Riemannian spaces of negative curvature*, Math. USSR Sbornik 25 (1975), 87–109.

[AB1] S.B. Alexander & R.L. Bishop, *The Hadamard-Cartan theorem in locally convex metric spaces*, L'Enseignement Math. 36 (1990), 309–320.

[AB2] S.B. Alexander & R.L. Bishop, *Comparison theorems for curves of bounded geodesic curvature in metric spaces of curvature bounded above*, Differential Geometry and Its Applications (to appear).

[ABB] S.B. Alexander, I.D. Berg & R.L. Bishop, *Geometric curvature bounds in Riemannian manifolds with boundary*, Trans. Amer. Math. Soc. 339 (1993), 703–716.

[Al1] A.D. Alexandrov, *A theorem on triangles in a metric space and some applications*, Trudy Math. Inst. Steklov 38 (1951), 5–23. (Russian; translated into German and combined with more material in [Ale2])

[Al2] A.D. Alexandrov, *Über eine Verallgemeinerung der Riemannschen Geometrie*, Schr. Forschungsinst. Math. Berlin 1 (1957), 33–84.

[Al3] A.D. Alexandrov, *Ruled surfaces in metric spaces*, Vestnik Leningrad Univ. 12 (1957), 5–26. (Russian)

[AlBe] A.D. Alexandrov & V.N. Berestovskiĭ, *Generalized Riemannian spaces*, Math. Encyclopaedia 4, 1984, pp. 1022–1026; English translation published by Kluver, Dordrecht

[AlBN] A.D. Alexandrov, V.N. Berestovskiĭ & I.G. Nikolaev, *Generalized Riemannian spaces*, Russian Math. Surveys 41 (1986), 1–54.

[AlRe] A.D. Alexandrov & Yu.G. Reshetnyak, *General Theory of Irregular Curves*, Kluwer Academic Publishers, Dordrecht, Boston, London, 1989.

[AlZa] A.D. Alexandrov & V.A. Zalgaller, *Intrinsic geometry of surfaces*, Translations of Math. Monographs 15, Amer. Math. Soc., Providence, Rhode Island, 1967.

[AloB] J.M. Alonso & M.R. Bridson, *Semihyperbolic groups*, Proceedings LMS (3) 70 (1995), 56–114.

[AnDG] F.D. Ancel, M.W. Davis & C.R.Guilbault, $CAT(0)$ *reflection manifolds*, Preprint, Ohio State University (1993).

[AnGu] F.D. Ancel & C.R. Guilbault, *Interiors of compact contractible n-manifolds are hyperbolic ($n \geq 5$)*, Preprint (1993).

[Anc1] A. Ancona, *Negatively curved manifolds, elliptic operators, and the Martin boundary*, Annals of Math. 125 (1987), 495–536.

[Anc2] A. Ancona, *Sur les fonctions propres positives des variétés de Cartan-Hadamard*, Comment. Math. Helvetici 64 (1989), 62–83.

[Anc3] A. Ancona, *Théorie du potentiel sur les graphes et les variétés*, Ecole d'Éte de Probabilités de Saint-Flour XVIII (P.L. Hennequin, ed.), Lecture Notes Maths. 1427, Springer-Verlag, 1990, pp. 5–112.

[Anc4] A. Ancona, *Convexity at infinity and Brownian motion on manifolds with unbounded negative curvature*, Revista Matemática Iberoamericana 10 (1994), 189–220.

[And] M. Anderson, *The Dirichlet problem at infinity for manifolds of negative curvature*, J. Differential Geometry 18 (1983), 701–721.

[AnSo] M. Anderson & R. Schoen, *Positive harmonic functions on complete manifolds of negative curvature*, Annals of Math. 121 (1985), 429–461.

[AnSr] M. Anderson & V. Schroeder, *Existence of flats in manifolds of nonpositive curvature*, Invent. math. 85 (1986), 303–315.

[Ano1] D.V. Anosov, *Geodesic flows on closed Riemannian manifolds with negative curvature*, Proc. Steklov Institute 90 (1967).

[Ano2] D.V. Anosov, *Tangent fields of transversal foliations in U-systems*, Math. Notes Acad. Sci. USSR 2:5 (1968), 818–823.

[AnSi] D.V. Anosov & Ya.G. Sinai, *Some smooth ergodic systems*, Russian Math. Surveys 22:5 (1967), 103–168.

[ArFa] C.S. Aravinda & F.T. Farrell, *Rank* 1 *aspherical manifolds which do not support any nonpositively curved metric*, Preprint 1993.

[Av] A. Avez, *Variétés Riemanniennes sans points focaux*, C. R. Acad. Sci. Paris 270 (1970), 188–191.

[AzW1] R. Azencott & E. Wilson, *Homogeneous manifolds with negative curvature, I*, Transactions Amer. Math. Soc. 215 (1976), 323–362.

[AzW2] R. Azencott & E. Wilson, *Homogeneous manifolds with negative curvature, II*, Memoirs Amer. Math. Soc. 178 (1976).

[Bab1] M. Babillot, *Théorie du renouvellement pour des chaines semi-markoviennes transientes*, Ann. Inst. Henri Poincaré 24 (1988), 507–569.

[Bab2] M. Babillot, *Comportement asymptotique du mouvement brownien sur une variété homogène à courbure sectionelle négative ou nulle*, Ann. Inst. Henri Poincaré 27 (1991), 61–90.

[Bab3] M. Babillot, *Potential at infinity and Martin boundary for symmetric spaces*, Harmonic analysis and discrete potential theory (Picardello, ed.), Proceedings Frascati Conference, Plenum Press, New York, 1992.

[Bab4] M. Babillot, *A probabilistic approach to heat diffusion on symmetric spaces*, J. Theoretical Probability 7 (1994), 599–607.

[Bab5] M. Babillot, *Asymptotics of Green functions on a class of solvable Lie groups*, Preprint, Paris VI (1994).

[Ba1] W. Ballmann, *Axial isometries of manifolds of nonpositive curvature*, Math. Annalen 259 (1982), 131–144.

[Ba2] W. Ballmann, *Nonpositively curved manifolds of higher rank*, Annals of Math. 122 (1985), 597–609.

[Ba3] W. Ballmann, *On the Dirichlet problem at infinity for manifolds of nonpositive curvature*, Forum Math. 1 (1989), 201–213.

[Ba4] W. Ballmann, *The Martin boundary of certain Hadamard manifolds*, Nieprint, Bonn 1990.

[Ba5] W. Ballmann, *Singular spaces of non-positive curvature*, Sur les groupes hyperboliques d'apres Mikhael Gromov (E. Ghys & P. de la Harpe, eds.), Progress in Math. 61, Birkhäuser, Boston-Basel-Berlin, 1990.

[BB1] W. Ballmann & M. Brin, *On the ergodicity of geodesic flows*, Ergodic Theory & Dynamical Systems 2 (1982), 311–315.

[BB2] W. Ballmann & M. Brin, *Polygonal complexes and combinatorial group theory*, Geometriae Dedicata 50 (1994), 165–191.

[BB3] W. Ballmann & M. Brin, *Orbihedra of nonpositive curvature*, SFB256-Preprint 367, Universität Bonn (1994).

[BBB1] W. Ballmann, M. Brin & K. Burns, *On surfaces with no conjugate points*, J. Differential Geometry 25 (1987), 249–273.

[BBB2] W. Ballmann, M. Brin & K. Burns, *On the differentiability of horocycles and horocyclic foliations*, J. Differential Geometry 26 (1987), 337–347.

[BBE] W. Ballmann, M. Brin & P. Eberlein, *Structure of manifolds of nonpositive curvature, I*, Annals of Math. 122 (1985), 171–203.

[BBS] W. Ballmann, M. Brin & R. Spatzier, *Structure of manifolds of nonpositive curvature, II*, Annals of Math. 122 (1985), 205–235.

[BaBu] W. Ballmann & S. Buyalo, *Nonpositively curved metrics on 2-polyhedra*, Math. Zeitschrift (to appear).

[BaEb] W. Ballmann & P. Eberlein, *Fundamental groups of manifolds of nonpositive curvature*, J. Differential Geometry 25 (1987), 1–22.

[BGS] W. Ballmann, M. Gromov & V. Schroeder, *Manifolds of nonpositive curvature* (1985), Birkhäuser, Boston-Basel-Stuttgart.

[BaL1] W. Ballmann & F. Ledrappier, *The Poisson boundary for rank one manifolds and their cocompact lattices*, Forum Math. 6 (1994), 301–313.

[BaL2] W. Ballmann & F. Ledrappier, *Discretization of positive harmonic functions on Riemannian manifolds and Martin boundary* (to appear).

[BaWo] W. Ballmann &M. Wojtkowski, *An estimate for the measure theoretic entropy of geodesic flows*, Ergodic Theory & Dynamical Systems 9 (1989), 271–279.

[BanS] V. Bangert & V. Schroeder, *Existence of flat tori in analytic manifolds of nonpositive curvature*, Annales Sci. Éc. Norm. Sup. 24 (1991), 605–634.

[Bar] S. Barré, *Polyèdres finis de dimension* 2 *à courbure* ≤ 0 *et de rang* 2, ENS, Lyon (1994).

[Ben1] N. Benakli, *Polyèdre à géometrie locale donnée*, C. R. Acad. Sci. Paris 313 (1991), 561–564.

[Ben2] N. Benakli, *Polyèdre hyperbolique à groupe d'automorphismes non discret*, C. R. Acad. Sci. Paris 313 (1991), 667–669.

[Ben3] N. Benakli, *Polyèdre avec un nombre fini de types de sommets*, C. R. Acad. Sci. Paris.

[Ben4] N. Benakli, *Polygonal complexes I: combinatorial and geometric properties*, Preprint, Princeton University (1993).

[BFL] Y. Benoist, P. Foulon & F. Labourie, *Flots d'Anosov à distributions stable et instable différentiables*, J. Amer. Math. Soc. 5 (1992), 33–74.

[Ber1] V. Berestovskiĭ, *On the problem of finite dimensionality of Busemann's G-space*, Sib. Mat. Z. 18 (1977), 219–221. (Russian)

[Ber2] V. Berestovskiĭ, *Borsuk's problem on the metrization of a polyhedron*, Dokl. Akad. Nauk CCCP 27 (1983), 273–277 (Russian); English translation in: Soviet Math. Doklady 27 (1983), 56–59.

[Ber3] V. Berestovskiĭ, *Spaces with bounded curvature and distance geometry*, Sib. Mat. Z. 27 (1986), 11–25 (Russian); English translation in: Siberian Math. J. 27 (1986), 8–19.

[Ber4] V. Berestovskiĭ, *Manifolds with intrinsic metric with one-sided bounded curvature in the sense of A.D. Alexandrov*, Math. Physics, Analysis, Geometry 1:1 (1994), 41–59. (Russian, summary in Ukranian)

[Ber5] V. Berestovskiĭ, *On Alexandrov's spaces with curvature bounded from above*, Russian Acad. Sci. Doklady (former Soviet Math. Dokl.) (to appear).

[BerN] V.N. Berestovskiĭ & I.G. Nikolaev, *Multidimensional generalized Riemannian spaces*, Geometry IV (Yu.G. Reshetnyak, ed.), Encyclopaedia of Math. Sciences 70, Springer-Verlag, Berlin-Heidelberg-New York..., 1993, pp. 165–243.

[Be] M. Berger, *Sur les groupes d'holonomie homogène des variétés à connexion affine et des variétés riemanniennes*, Bull. Soc. Math. France 83 (1953), 279–330.

[BCG] G. Besson, G. Courtois & S. Gallot, *Volumes, entropies et rigidités des espaces localement symétriques de courbure strictement négative*, C. R. Acad. Sci. Paris 319 (1994), 81–84.

[BiGe] R. Bieri & R. Geoghegan, *Kernels of actions on nonpositively curved spaces*, Preprint, Binghamton University (1995).

[BiON] R. Bishop & B. O'Neill, *Manifolds of negative curvature*, Transactions Amer. Math. Soc. 145 (1969), 1–49.

[Bl] L.M. Blumenthal, *Theory and applications of distance geometry*, Oxford at the Clarendon Press, 1953.

[BlMe] L.M. Blumenthal & K. Menger, *Studies in geometry*, W.H. Freeman and Company, San Francisco, 1970.

[Bor1] A. Borbély, *A note on the Dirichlet problem at infinity for manifolds of negative curvature*, Proceedings Amer. Math. Soc. 114 (1992), 865–872.

[Bor2] A. Borbély, *On convex hulls in negatively curved manifolds*, Preprint.

[Bow] B.H. Bowditch, *A class of incomplete non-positively curved manifolds*, Preprint, University of Melbourne (1991).

[Bri1] M.R. Bridson, *Geodesics and curvature in metric simplicial complexes*, Group theory from a geometrical viewpoint (E. Ghys, A. Haefliger & A. Verjovsky, eds.), Proceedings ICTP, Trieste, World Scientific, Singapore, 1991.

[Bri2] M.R. Bridson, *On the existence of flat planes in spaces of nonpositive curvature*, Proceedings Amer. Math. Soc. (to appear).

[BrHa] M.R. Bridson & A. Haefliger, *Spaces of nonpositive curvature*, In preparation.

[BrK1] M. Brin & Y. Kifer, *Harmonic measures on covers of compact surfaces of nonpositive curvature*, Transactions Amer. Math. Soc. 340 (1993), 373–393.

[BrK2] M. Brin & Y. Kifer, *Brownian motion and harmonic functions on polygonal complexes*, Preprint (1993).

[Bro] K.S. Brown, *Buildings*, Springer-Verlag, Berlin-Heidelberg-New York, 1989.

[BruT] F. Bruhat & J. Tits, *Groupes réductifs sur un corps local, I. Données radicielles valuées*, Publications Math. IHES 41 (1972), 5–251.

[BurS] M. Burger & V. Schroeder, *Amenable groups and stabilizers of measures on the boundary of a Hadamard manifold*, Math. Ann. 276 (1987), 505–514.

[Bu] K. Burns, *Hyperbolic behaviour of geodesic flows on manifolds with no focal points*, Ergodic Theory & Dynamical Systems 3 (1983), 1–12.

[BuKa] K. Burns & A. Katok, *Manifolds of non-positive curvature*, Ergodic Theory & Dynamical Systems 5 (1985), 307–317.

[BuSp] K. Burns & R. Spatzier, *Manifolds of nonpositive curvature and their buildings*, Publications Math. IHES 65 (1987), 35–59.

[Bus1] H. Busemann, *Spaces with non-positive curvature*, Acta Mathematica 80 (1948), 259–310.

[Bus2] H. Busemann, *The geometry of geodesics*, Academic Press Inc., New York, 1955.

[BuKa] P. Buser & H. Karcher, *Gromov's almost flat manifolds*, Astérisque 81, Soc. Math. France, 1981.

[Buy1] S. Buyalo, *Closed geodesics on two-dimensional orbifolds of nonpositive curvature*, Leningrad Math. J. 1 (1990), 653–673.

[Buy2] S. Buyalo, *Collapsing manifolds of nonpositive curvature. I*, Leningrad Math. J. 1 (1990), 1135–1155.

[Buy3] S. Buyalo, *Collapsing manifolds of nonpositive curvature. II*, Leningrad Math. J. 1 (1990), 1371–1399.

[Buy4] S. Buyalo, *Euclidean planes in open three-dimensional manifolds of nonpositive curvature*, St. Petersburg Math. J. 3 (1992), 83–96.

[Buy5] S. Buyalo, *Homotopy invariance of some geometric properties of three-dimensional manifolds of nonpositive curvature*, St. Petersburg Math. J. 3 (1992 791–808).

[Buy6] S. Buyalo, *An example of a negatively curved 4-manifold*, St.Petersburg Math. J.5 (1994), 171–176.

[Buy7] S. Buyalo, *The finiteness theorem for three-dimensional manifolds of nonpositive curvature*, Amer. Math. Soc. Translations 159 (1994), 25–43.

[BuK1] S. Buyalo & V. Kobel'skii, *Cusp closing of hyperbolic manifolds*, Geometriae Dedicata (to appear).

[BuK2] S. Buyalo & V. Kobel'skii, *Geometrization of graph-manifolds: conformal states*, St. Petersburg Math. J. (to appear).

[BuK3] S. Buyalo & V. Kobel'skii, *Geometrization of graph-manifolds: isometric states*, St. Petersburg Math. J. (to appear).

[Ca] E. Cartan, *Lecons sur la géométrie des espaces de Riemann*, Gauthier-Villars, Paris, 1928.

[ChD1] R. Charney & M.W. Davis, *Singular metrics of nonpositive curvature on branched covers of Riemannian manifolds*, Amer. J. of Math. 115 (1993), 929–1009.

[ChD2] R. Charney & M.W. Davis, *The Euler characteristic of a nonpositively curved, piecewise Euclidean manifold*, Pacific J. Math. (to appear).

[ChD3] R. Charney & M.W. Davis, *The polar dual of a convex polyhedral set in hyperbolic space*, Preprint, Ohio State University.

[ChEb] J. Cheeger & D. Ebin, *Comparison theorems in Riemannian geometry*, North-Holland/ American Elsevier (1975).

[Che1] S.-S. Chen, *Complete homogeneous Riemannian manifolds of negative sectional curvature*, Comment. Math. Helv. 50 (1975), 115–122.

[Che2] S.-S. Chen, *Duality condition and Property* (*S*), Pacific J. Math. 98 (1982), 313–322.

[CE1] S.-S. Chen & P. Eberlein, *Isometry groups of simply connected manifolds of nonpositive curvature*, Illinois Journal of Math. 24 (1980), 73–103.

[CE2] S.-S. Chen & P. Eberlein, *Isometry classes of lattices of nonpositive curvature and uniformly bounded volume*, Bol. Soc. Brasil Mat. 13 (1982), 25–44.

[Coh] S. Cohn-Vossen, *Existenz kürzester Wege*, Doklady SSSR 8 (1935), 339–342.

[Co1] K. Corlette, *Flat G-bundles with canonical metrics*, J. Differential Geometry 28 (1988), 361–382.

[Co2] K. Corlette, *Archimedean superrigidity and hyperbolic geometry*, Annals of Math. 135 (1990), 165–182.

[Cor] V. Cortés Suárez, *Alekseevskiĭs quaternionische Kählermannigfaltigkeiten*, Dissertation, Universität Bonn (1993).

[Cro] C. Croke, *Rigidity for surfaces of non-positive curvature*, Comment. Math. Helv. 65 (1990), 150–169.

[CEK] C. Croke, P. Eberlein & B Kleiner, *Conjugacy and rigidity for nonpositively curved manifolds of higher rank*, Preprint.

[DaRi] E. Damek & F. Ricci, *A class of non-symmetric harmonic Riemannian spaces*, Bulletin Amer. Math. Soc. 27 (1992), 139–142.

[DADM] J. D'Atri & I. Dotti Miatello, *A characterization of bounded symmetric domains by curvature*, Transactions Amer. Math. Soc. 276 (1983), 531–540.

[Dav] M.W. Davis, *Nonpositive curvature and reflection groups*, Preprint, Ohio State University (1994).

[DaJa] M.W. Davis & T. Januszkiewicz, *Hyperbolization of polyhedra*, J. Differential Geometry 34 (1991), 347–388.

[Dr1] M. Druetta, *Homogeneous Riemannian manifolds and the visibility axiom*, Geometriae Dedicata 17 (1985), 239–251.

[Dr2] M. Druetta, *Visibility and rank one in homogeneous spaces of* $K \leq 0$, Transactions Amer. Math. Soc. 304 (1987), 307–321.

[Dr3] M. Druetta, *The rank in homogeneous spaces of nonpositive curvature*, Proceedings Amer. Math. Soc. 105 (1989), 972–979.

[Dr4] M. Druetta, *Fixed points of isometries at infinity in homogeneous spaces*, Illinois J. Math. 33 (1989), 210–226.

[Dr5] M. Druetta, *Nonpositively curved homogeneous spaces of dimension five*, Pacific J. Math. 148 (1991), 17–37.

[Dr6] M. Druetta, *The Lie algebra of isometries of homogeneous spaces of nonpositive curvature*, Geometriae Dedicata (to appear).

[Dy] E.B. Dynkin, *Brownian motion in certain symmetric spaces and nonnegative eigenfunctions of the Laplace-Beltrami operator*, Izv. Akad. Nauk. SSSR 30 (1966), 455–478 (Russian); English translation in: Amer. Math. Soc. Translations 72 (1968), 203–228.

[Eb1] P. Eberlein, *Geodesic flow in certain manifolds without conjugate points*, Transactions Amer. Math. Soc. 167 (1972), 151-170.

[Eb2] P. Eberlein, *Geodesic flows on negatively curved manifolds, I*, Annals of Math. 95 (1972), 492–510.

[Eb3] P. Eberlein, *Geodesic flows on negatively curved manifolds, II*, Transactions Amer. Math. Soc. 178 (1973), 57–82.

[Eb4] P. Eberlein, *When is a geodesic flow of Anosov type? I*, J. Differential Geometry 8 (1973), 437–463.

[Eb5] P. Eberlein, *When is a geodesic flow of Anosov type? II*, J. Differential Geometry 8 (1973), 565–577.

[Eb6] P. Eberlein, *Some properties of the fundamental group of a Fuchsian manifold*, Inventiones math. 19 (1973), 5–13.

[Eb7] P. Eberlein, *Surfaces of nonpositive curvature*, Memoirs Amer. Math. Soc. 218 (1979), 1–90.

[Eb8] P. Eberlein, *Lattices in manifolds of nonpositive curvature*, Annals of Math. 111 (1980), 435–476.

[Eb9] P. Eberlein, *Geodesic rigidity in nonpositively curved manifolds*, Transactions Amer. Math. Soc. 268 (1981), 411–443.

[Eb10] P. Eberlein, *Isometry groups of simply connected manifolds of nonpositive curvature, II*, Acta Math. 149 (1982), 41–69.

[Eb11] P. Eberlein, *A canonical form for compact nonpositively curved manifolds whose fundamental groups have nontrivial center*, Math. Ann. 260 (1982), 23–29.

[Eb12] P. Eberlein, *Rigidity of lattices of nonpositive curvature*, Ergodic Theory & Dynamical Systems 3 (1983), 47–85.

[Eb13] P. Eberlein, *The Euclidean de Rham factor of a lattice of nonpositive curvature*, J. Differential Geometry 18 (1983), 209–220.

[Eb14] P. Eberlein, *Structure of manifolds of nonpositive curvature*, Lecture Notes in Mathematics 1156, Springer-Verlag, Berlin-Heidelberg-New York yr 1985, pp. 86–153.

[Eb15] P. Eberlein, *L-subgroups in spaces of nonpositive curvature*, Lecture Notes in Mathematics 1201, Springer-Verlag, Berlin-Heidelberg-New York, 1986, pp. 41–88.

[Eb16] P. Eberlein, *Symmetry diffeomorphism group of a manifold of nonpositive curvature, I*, Transactions Amer. Math. Soc. 309 (1988), 355–374.

[Eb17] P. Eberlein, *Symmetry diffeomorphism group of a manifold of nonpositive curvature, II*, Indiana Univ. Math. J. 37 (1988), 735–752.

[Eb18] P. Eberlein, *Manifolds of nonpositive curvature*, Global Differential Geometry (S.S. Chern, ed.), MAA Studies in Mathematics 27, 1989, pp. 223–258.

[Eb19] P. Eberlein, *Geometry of nonpositively curved manifolds*, In preparation, 1995.

[EbHS] P. Eberlein, U. Hamenstädt & V. Schroeder, *Manifolds of nonpositive curvature*, Proceedings Symp. Pure Math. 54, Part 3 (1993), 179–227.

[EbHe] P. Eberlein & J. Heber, *A differential geometric characterization of symmetric spaces of higher rank*, Publications Math. IHES 71 (1990), 33–44.

[EbON] P. Eberlein & B. O'Neill, *Visibility manifolds*, Pacific J. Math. 46 (1973), 45–109.

[EeSa] J. Eells & J. Sampson, *Harmonic maps of Riemannian manifolds*, Amer. J. Math. 86 (1964), 109–160.

[EsSc] J. Eschenburg & V. Schroeder, *Tits distance of Hadamard manifolds and isoparametric hypersurfaces*, Geometriae Dedicata 40 (1991), 97–101.

[FaJ1] F.T. Farrel & L.E. Jones, *A topological analogue of Mostow's rigidity theorem*, J. Amer. Math. Soc. 2 (1989), 257–370.

[FaJ2] F.T. Farrel & L.E. Jones, *Negatively curved manifolds with exotic smooth structures*, J. Amer. Math. Soc. 2 (1989), 899–908.

[FaJ3] F.T. Farrel & L.E. Jones, *Classical aspherical manifolds*, CBMS regional conference 75, Amer. Math. Soc., 1990.

[FaJ4] F.T. Farrel & L.E. Jones, *Stable pseudoisotopy spaces of compact nonpositively curved manifolds*, J. Differential Geometry 34 (1991), 769–834.

[FaJ5] F.T. Farrel & L.E. Jones, *Nonuniform hyperbolic lattices and exotic smooth structures*, J. Differential Geometry 38 (1993), 235–261.

[FaJ6] F.T. Farrel & L.E. Jones, *Topological rigidity for compact nonpositively curved manifolds*, Proceedings Symp. Pure Math. 54 III (1993), 229–274.

[FaJ7] F.T. Farrel & L.E. Jones, *Complex hyperbolic manifolds and exotic smooth structures*, Inventiones math. 117 (1994), 57–74.

[Fe] R. Feres, *Geodesic flows on manifolds of negative curvature with smooth horospheric foliations*, Ergodic Theory & Dynamical Systems 11 (1991), 653–686.

[FeKa] R. Feres & A. Katok, *Anosov flows with smooth foliations and three-dimensional manifolds of negative curvature*, Ergodic Theory & Dynamical Systems 10 (1990), 657–670.

[FoSc] S. Fornari & V. Schroeder, *Ramified coverings with nonpositive curvature*, Math. Z. 203 (1990), 123–128.

[FoLa] P. Foulon & F. Labourie, *Sur les variétés compactes asymptotiquement harmoniques*, Inventiones math. 109 (1992), 97–111.

[FrMa] A. Freire & R. Mañé, *On the entropy of the geodesic flow in manifolds without conjugate points*, Inventiones math. 69 (1982), 375–392.

[Fu1] H. Furstenberg, *A Poisson formula for semi-simple Lie groups*, Annals of Math. 77 (1963), 335–386.

[Fu2] H. Furstenberg, *Random walks and discrete subgroups of Lie groups*, Advances in Probability and Related Topics, I (P. Ney, ed.), Dekker, New York, 1971, pp. 1–63.

[Fu3] H. Furstenberg, *Boundary theory and stochastic processes on homogeneous spaces*, Proceedings Symp. Pure Math. 26, 1973, pp. 193–229.

[GeSh] S. Gersten & H. Short, *Small cancellation theory and automatic groups*, Inventiones math. 102 (1990), 305–334.

[Gh] E. Ghys, *Flots d'Anosov dont les feuilletages stables sont différentiables*, Ann. Sci. Ecole Norm. Sup. 20 (1987), 251–270.

[GhHa] E. Ghys & P. de la Harpe, *Sur les groupes hyperboliques d'après Mikhael Gromov*, Progress in Math. 83, Birkhäuser, Boston-Basel-Stuttgart, 1990.

[GoSc] S. Goette & V. Schroeder, *Totally geodesic hypersurfaces in manifolds of nonpositive curvature*, Manuscripta Math. (to appear).

[GroW] D. Gromoll & J. Wolf, *Some relations between the metric structure and the algebraic structure of the fundamental group in manifolds of nonpositive curvature*, Bulletin Amer. Math. Soc. 77 (1971), 545–552.

[Gr1] M. Gromov, *Manifolds of negative curvature*, J. Differential Geometry 13 (1978), 223–230.

[Gr2] M. Gromov, *Synthetic geometry in Riemannian manifolds*, Proceedings of the International Congress of Mathematicians, Helsinki (O. Lehto, ed.), 1978, vol. 1, Adacemia Scientarium Fennica, 1980, pp. 415–420.

[Gr3] M. Gromov, *Hyperbolic manifolds, groups and actions*, Riemann Surfaces and Related Topics (I. Kra and B. Maskit, eds.), Proceedings, Stony Brook 1978, Annals of Math. Studies 97, Princeton University, 1981, pp. 83–213.

[Gr4] M. Gromov, *Structures métriques pour les variétés Riemanniennes*, rédigé par J. Lafontaine et P. Pansu, Cedic/Fernand Nathan, 1981.

[Gr5] M. Gromov, *Hyperbolic groups*, Essays in group theory (S.M. Gersten, ed.), MSRI Publ. 8, Springer-Verlag, Berlin-Heidelberg-New York, 1987, pp. 75–264.

[Gr6] M. Gromov, *Asymptotic invariants of infinite groups*, Geometric Group Theory, vol.2 (G. A. Noble and M. A. Roller, eds.), London Math. Soc. Lecture Notes 182, 1993.

[GrPa] M. Gromov & P. Pansu, *Rigidity of lattices: an introduction*, Geometric topology: recent developments, Montecatini, 1990, pp. 39–137.

[GrPi] M. Gromov & I. Piatetski-Shapiro, *Non-arithmetic groups in Lobachevsky spaces*, Publications Math. IHES 66 (1988), 93–103.

[GrSc] M. Gromov & R. Schoen, *Harmonic maps into singular spaces and p-adic superrigidity for lattices in groups of rank one*, Publications Math. 76 (1992), 165–246.

[GrTh] M. Gromov & W. Thurston, *Pinching constants for hyperbolic manifolds*, Inventiones math. 89 (1987), 1–12.

[Gu] Y. Guivarc'h, *Marches aléatoires sur les groupes et problèmes connexes*, Preprint, Rennes.

[GuJT] Y. Guivarc'h, L. Ji & J.C. Taylor, *The Martin compactification of symmetric spaces.*

[Had] J. Hadamard, *Les surfaces à courbure opposées et leurs lignes géodésiques*, J. Math. Pures Appl. 4 (1898), 27–73.

[Hae] A. Haefliger, *Complexes of groups and orbihedra*, Group theory from a geometric viewpoint (E. Ghys, A. Haefliger & A. Verjovski, eds.), World Scientific, Singapore, 1991, pp. 504–540.

[Hag] F. Haglund, *Les polyèdres de Gromov*, C. R. Acad. Sci. Paris 313 (1991), 603–606.

[Ham1] U. Hamenstädt, *Regularity at infinity of compact negatively curved manifolds*, Ergodic Theory & Dynamical Systems 14 (1994).

[Ham2] U. Hamenstädt, *Harmonic measures for leafwise elliptic operators*, First European Congress of Mathematics (Paris, 1992), II, Progress in Mathematics 121, Birkhäuser Verlag, Basel-Boston-Berlin, 1994, pp. 73–96.

[Ham3] U. Hamenstädt, *Invariant two-forms for geodesic flows*, Math. Ann. (to appear).

[Ham4] U. Hamenstädt, *Harmonic measures for compact negatively curved manifolds*, Preprint, Universität Bonn (1995).

[Har] P. Hartman, *On homotopic harmonic maps*, Canadian J. Math. 19 (1967), 673–687.

[Heb] J. Heber, *On the geometric rank of homogeneous spaces of nonpositive curvature*, Invent. math. 112 (1993), 151–170.

[He1] E. Heintze, *On homogeneous manifolds of negative curvature*, Math. Ann. 211 (1974), 23–34.

[He2] E. Heintze, *Compact quotients of homogeneous negatively curved Riemannian manifolds*, Math. Z. 140 (1974), 79–80.

[He3] E. Heintze, *Mannigfaltigkeiten negativer Krümmung*, Habilitationsschrift, Bonn 1976.

[HeIH] E. Heintze & H.C. Im Hof, *Geometry of horospheres*, J. Differential Geometry 12 (1977), 481–491.

[Hel] S. Helgason, *Differential geometry, Lie groups and symmetric spaces*, Academic Press, 1978.

[Her] R. Hermann, *Homogeneous Riemannian manifolds of nonpositive sectional curvature*, Nederl. Akad. Wetensch. Proc. 25 (1963), 47–56.

[HiPu] M. Hirsch & C. Pugh, *Smoothness of horocycle foliations*, J. Differential Geometry 10 (1975), 225–238.

[Ho1] E. Hopf, *Statistik der geodätischen Linien in Mannigfaltigkeiten negativer Krümmung*, Berichte Verh. Sächs. Akad. Wiss. Leipzig 91 (1939), 261–304.

[Ho2] E. Hopf, *Statistik der Lösungen geodätischer Probleme vom unstabilen Typus. II*, Math. Ann. 117 (1940), 590–608.

[Hor] P. Horja, *On the number of geodesic segments connecting two points on manifolds of non-positive curvature*, Preprint, Duke University (1995).

[HumS] C. Hummel & V. Schroeder, *Cusp closing in rank one symmetric spaces*, Preprint.

[Hu] B. Hu, *Whitehead groups of finite polyhedra with nonpositive curvature*, J. Differential Geometry 38 (1993), 501–517.

[HurK] S. Hurder and A. Katok, *Differentiability, rigidity and Godbillon-Vey classes for Anosov flows*, Publications Math. IHES 72 (1990), 5–61.

[IH1] H. C. Im Hof, *The family of horospheres through two points*, Math. Ann. 240 (1979), 1–11.

[IH2] H. C. Im Hof, *Die Geometrie der Weylkammern in symmetrischen Räumen vom nichtkompakten Typ*, Habilitationsschrift, Universität Bonn (1979).

[IH3] H. C. Im Hof, *An Anosov action on the bundle of Weyl chambres*, Ergodic Theory & Dynamical Systems 5 (1985), 587–593.

[Jo1] J. Jost, *Equilibrium maps between metric spaces*, Calc. Variations 2 (1994), 173–204.

[Jo2] J. Jost, *Convex functionals and generalized harmonic maps into spaces of nonpositive curvature*, Preprint (1993).

[JoYa] J. Jost & S.-T. Yau, *Harmonic maps and superrigidity*, Proc. Symp. Pure Math. 54 (1993), 245–280.

[JoZu] J. Jost & K. Zuo, *Harmonic maps of infinite energy and rigidity results for quasiprojective varieties*, Preprint, Universit́at Bochum (1994).

[Kai1] V.A. Kaimanovich, *An entropy criterion for maximality of the boundary of random walks on discrete groups*, Doklady AN SSSR 280 (1985), 1051–1054 (Russian); English translation: Soviet Math. Dokl. 31 (1985), 193–197.

[Kai2] V.A. Kaimanovich, *Discretization of bounded harmonic functions on Riemannian manifolds and entropy*, Proceedings of the International Conference on Potential Theory, Nagoya (M. Kishi, ed.), De Gruyter, Berlin, 1992, pp. 212–223.

[Kai3] V.A. Kaimanovich, *The Poisson boundary of hyperbolic groups*, C. R. Acad. Sci. Paris 318 (1994), 59–64.

[Kan] M. Kanai, *Geodesic flows of negatively curved manifolds with smooth stable and unstable foliations*, Ergodic Theory & Dynamical Systems 8 (1988), 215–239.

[KaL1] M. Kapovich & B. Leeb, *On asymptotic cones and quasi-isometry classes of fundamental groups of 3-manifolds*, GAFA, (to appear).

[KaL2] M. Kapovich & B. Leeb, *On actions of discrete groups on nonpositively curved spaces*, MSRI-Preprint 059-94, 1994.

[KaL3] M. Kapovich & B. Leeb, *On quasi-isometries of graph-manifold groups*, Preprint 1994.

[Kar] H. Karcher, *Riemannian comparison constructions*, Global differential geometry (S.-S. Chern, ed.), MAA Studies in Math. 27, 1989, pp. 170–222.

[Karp] F. I. Karpelevič, *The geometry of geodesics and the eigenfunctions of the Beltrami-Laplace operator on symmetric spaces*, Amer. Math. Soc. Translations 14 (1965), 51–199.

[Kat1] A. Katok, *Entropy and closed geodesics*, Ergodic Theory & Dynamical Systems 2 (1982), 339–365.

[Kat2] A. Katok, *Four applications of conformal equivalence to geometry and dynamics*, Ergodic Theory & Dynamical Systems 8 (1988), 139–152.

[Ke] W. S. Kendall, *Probability, convexity, and harmonic maps II: smoothness via probabilistic gradient inequalities*, Preprint, University of Warwick (1993).

[Ki1] Y. Kifer, *Brownian motion and harmonic functions on manifolds of negative curvature*, Theory Prob. Appl. 21 (1976), 81–95.

[Ki2] Y. Kifer, *Brownian motion and harmonic functions on complete manifolds of nonpositive curvature*, Pitman Research Notes in Mathematics 150 (1986), 132–178.

[Kl] B. Kleiner, *A metric characterization of Tits buildings*, To be in preparation (1995).

[KlLe] B. Kleiner & B. Leeb, *Rigidity of quasi-isometries for symmetric spaces of higher rank*, Preprint, Bonn January 1995.

[Kn] G. Knieper, *Das Wachstum der Äquivalenzklassen geschlossener Geodätischen in kompakten Mannigfaltigkeiten*, Archiv Math. 40 (1983), 559–568.

[Kob] S. Kobayashi, *Homogeneous Riemannian manifolds of negative curvature*, Bulletin Amer. Math. Soc 68 (1962), 338–339.

[KoSc] N. Korevaar & R. Schoen, *Sobolev spaces and harmonic maps for metric space targets*, Communications in Analysis and Geometry 1 (1993), 561–659.

[Lab] F. Labourie, *Existence d'applications harmoniques tordues à valeurs dans les variétés à courbure négative*, Proc. Amer. Math. Soc. 111 (1991), 877–882.

[LaYa] H. B. Lawson & S. T. Yau, *On compact manifolds of nonpositive curvature*, J. Differential Geometry 7 (1972), 211–228.

[Le] B. Leeb, 3-*manifolds with(out) metrics of nonpositive curvatures*, Inventiones Math. (to appear) (1992).

[Leu1] E. Leuzinger, *Une relation fondamentale pour les triangles dans les espaces symétriques*, C. R. Acad. Sci. Paris 312 (1991), 451–454.

[Leu2] E. Leuzinger, *Sur une nouvelle classe d'actions d'Anosov*, C. R. Acad. Sci. Paris 315 (1992), 65–68.

[Leu3] E. Leuzinger, *On the trigonometry of symmetric spaces*, Comment. Math. Helvetici 67 (1992), 252–286.

[Leu4] E. Leuzinger, *New geometric examples of Anosov actions*, Annals of Global Analysis and Geometry 12 (1994), 173–181.

[Leu5] E. Leuzinger, *Geodesic rays in locally symmetric spaces*, Differential Geometry and Applications (to appear).

[Leu6] E. Leuzinger, *An exhaustion of locally symmetric spaces by compact submanifolds with corners*, Preprint, Universität Zürich (1994).

[Lo] J. Lohkamp, *An existence theorem for harmonic maps*, Manuscripta math. 67 (1990), 21–23.

[LySc] R.C. Lyndon & P.E. Schupp, *Combinatorial group theory*, Ergebnisse der Math. 89, Springer-Verlag, Berlin-Heidelberg-New York, 1977.

[LySu] T. Lyons & D. Sullivan, *Function theory, random paths and covering spaces*, J. Differential Geometry 19 (1984), 299–323.

[Man1] A. Manning, *Topological entropy for geodesic flows*, Annals of Math. 110 (1979), 567–573.

[Man2] A. Manning, *Curvature bounds for the entropy of the geodesic flow on a surface*, J. London Math. Soc. 24 (1981), 351–357.

[Mar1] G.A. Margulis, *Discrete groups of motion of manifolds of nonpositive curvature*, Proceedings Int. Congress Math. (Vancouver 1974), Amer. Math. Soc. Translations 109 (1977), 33–45.

[Mar2] G.A. Margulis, *Discrete subgroups of semisimple Lie groups*, Ergebnisse der Mathematik und ihrer Grenzgebiete 3/17, Springer-Verlag, Berlin-Heidelberg-New York, 1990.

[Mar3] G.A. Margulis, *Superrigidity for commensurability subgroups and generalized harmonic maps*, Preprint (1994).

[Ma] F. I. Mautner, *Geodesic flows on symmetric Riemannian spaces*, Annals of Math. 65 (1957), 416–431.

[Me] K. Menger, *The formative years of Abraham Wald and his work in geometry*, Annals of Math. Statistics 23 (1952), 14–20.

[MSY] N. Mok, Y.-T. Siu & S.-K. Yeung, *Geometric superrigidity*, Inventiones math. 113 (1993), 57–83.

[Mor] M. Morse, *A fundamental class of geodesics on any surface of genus greater than one*, Transactions Amer. Math. Soc. 26 (1924), 25–60.

[Mos] G.D. Mostow, *Strong rigidity of locally symmetric spaces*, Annals of Math. Studies 78, Princeton University Press, 1973.

[MoSi] G.D. Mostow & Y.-S. Siu, *A compact Kähler surface of negative curvature not covered by the ball*, Annals of Math. 112 (1980), 321–360.

[Mou] G. Moussong, *Hyperbolic Coxeter groups*, Ph.D. thesis, Ohio State University (1988).

[Ni1] I.G. Nikolaev, *Space of directions at a point of a space of curvature not greater than K*, Siberian Math. J. 19 (1978), 944–949.

[Ni2] I.G. Nikolaev, *Solution of the Plateau problem in spaces of curvature at most K*, Siberian Math. J. 20 (1979), 246–252.

[Ni3] I.G. Nikolaev, *On the axioms of Riemannian geometry*, Soviet Math Doklady 40 (1990), 172–174.

[Ni4] I.G. Nikolaev, *Synthetic methods in Riemannian geometry*, University of Illinois, Urbana-Champaign, 1992.

[Ni5] I.G. Nikolaev, *The tangent cone of an Alexandrov space of curvature $\leq K$*, Manuscripta Math. (to appear).

[Ol] M. A. Olshanetskii, *Martin boundary for the Laplace-Beltrami operator on a Riemannian symmetric space of nonpositive curvature*, Uspeki Math. Nauk. 24:6 (1969), 189–190.

[OsSa] R. Ossermann & P. Sarnak, *A new curvature invariant and entropy of geodesic flows*, Inventiones math. 77 (1984), 455–462.

[Pe1] Ya. B. Pesin, *Characteristic Lyapunov exponents and smooth ergodic theory*, Russian Math. Surveys 32 (1977), 55–114.

[Pe2] Ya. B. Pesin, *Geodesic flow on closed Riemannian manifolds without focal points*, Math. USSR Izvestija 11 (1977), 1195–1228.

[Pe3] Ya B. Pesin, *Equations for the entropy of a geodesic flow on a compact Riemannian manifold without conjugate points*, Math. Notes 24 (1978), 796–805.

[PrRa] G. Prasad & M.S. Raghunathan, *Cartan subgroups and lattices in semisimple groups*, Annals of Math. 96 (1972), 296–317.

[Pr] A. Preissmann, *Quelques propriétés globales des espaces de Riemann*, Comment. Math. Helv. 15 (1943), 175–216.

[PuSh] C. Pugh & M. Shub, *Ergodic Attractors*, Transactions Amer. Math. Soc. 312 (1989), 1–54.

[Re1] Yu.G. Reshetnyak, *On the theory of spaces of curvature not greater than K*, Mat. Sbornik 52 (1960), 789–798. (Russian)

[Re2] Yu.G. Reshetnyak, *Nonexpanding maps in spaces of curvature not greater than K*, Sib. Mat. Z. 9 (1968), 918–928 (Russian); English translation published in 1969 with title: *Inextensible mappings in a space of curvature no greater than K*, Siberian Math. J. 9 (1968), 683–689.

[Re3] Yu.G. Reshetnyak, *Two-dimensional manifolds of bounded curvature*, Geometry IV, Non-regular Riemannian Geometry (Yu.G.Reshetnyak, ed.), Encyclopaedia of Mathematical Sciences 70, Springer-Verlag, Berlin-Heidelberg-New York ..., 1993, pp. 3–163.

[Ri] W. Rinow, *Die innere Geometrie der metrischen Räume*, Grundlehren der math. Wissenschaften 105, Springer-Verlag, Berlin-Göttingen-Heidelberg, 1961.

[Sar] P. Sarnak, *Entropy estimates for geodesic flows*, Ergodic Theory & Dynamical Systems 2 (1982), 513–524.

[Sas] S. Sasaki, *On the differential geometry of tangent bundles of Riemannian manifolds*, Tôhoku J. Math. 10 (1958), 338–354.

[Sch1] V. Schroeder, *Über die Fundamentalgruppe von Räumen nichtpositiver Krümmung mit endlichem Volumen*, Schriftenreihe des Mathematischen Instituts Münster, 1984.

[Sch2] V. Schroeder, *Finite volume and fundamental group on manifolds of negative curvature*, J. Differential Geometry 20 (1984), 175–183.

[Sch3] V. Schroeder, *A splitting theorem for spaces of nonpositive curvature*, Inventiones math. 79 (1985), 323–327.

[Sch4] V. Schroeder, *On the fundamental group of a visibility manifold*, Math. Z. 192 (1986), 347–351.

[Sch5] V. Schroeder, *Rigidity of nonpositively curved graphmanifolds*, Math. Ann. 274 (1986), 19–26.

[Sch6] V. Schroeder, *Flache Unterräume in Mannifaltigkeiten nichtpositiver Krümmung*, Habilitationsschrift, Universität Münster 1987.

[Sch7] V. Schroeder, *Existence of immersed tori in manifolds of nonpositive curvature*, J. reine angew. Mathematik 390 (1988), 32–46.

[Sch8] V. Schroeder, *Structure of flat subspaces in low-dimensional manifolds of nonpositive curvature*, Manuscripta Math. 64 (1989), 77–105.

[Sch9] V. Schroeder, *A cusp closing theorem*, Proceedings Amer. Math. Soc. 106 (1989), 797–802.

[Sch10] V. Schroeder, *Codimension one tori in manifolds of nonpositive curvature*, Geometriae Dedicata 33 (1990), 251–263.

[Sch11] V. Schroeder, *Analytic manifolds of nonpositve curvature with higher rank subspaces*, Archiv. Math. 56 (1991), 81–85.

[ScSt] V. Schroeder & M. Strake, *Local rigidity of symmetric spaces with nonpositive curvature*, Proceedings Amer. Math. Soc. 106 (1989), 481–487.

[ScZi] V. Schroeder & W. Ziller, *Local rigidity of symmetric spaces*, Transactions Amer. Math. Soc. 320 (1990), 145–160.

[Schw] R.E. Schwartz, *The quasi-isometry classification of the Hilbert modular group*, Preprint (1995).

[She1] S.Z. Shefel, *On the intrinsic geometry of saddle surfaces*, Sib. Mat. Z. 5 (1964), 1382–1396. (Russian)

[She2] S.Z. Shefel, *Geometric properties of embedded manifolds*, Sib. Mat. Z. 26 (1985), 170–188 (Russian); English translation in: Siberian Math. J. 26 (1985), 133–147.

[Shi] K. Shiga, *Hadamard manifolds*, Advanced Studies in Pure Math. 3 (1984), 239–281.

[Si] J. Simons, *On the transitivity of holonomy systems*, Annals of Math. 76 (1962), 213–234.

[Sp1] R. Spatzier, *Dynamical properties of algebraic systems*, Dissertation, Univ. of Warwick (1983).

[Sp2] R. Spatzier, *Harmonic analysis in rigidity theory*, Preprint, University of Michigan, Ann Arbor (1994).

[Su] D. Sullivan, *The Dirichlet problem at infinity for a negatively curved manifold*, J. Differential Geometry 18 (1983), 723–732.

[Sw] J. Świątkowski, *Polygonal complexes of nonpositive curvature: from local symmetry to global one*, Preprint, Wroclaw 1993.

[Ti1] J. Tits, *Free subgroups in linear groups*, J. Algebra 20 (1972), 250–270.

[Ti2] J. Tits, *Buildings of spherical type and finite BN-pairs*, Lecture Notes in Math. 386, Springer-Verlag, Berlin-Heidelberg-New York, 1974.

[Thu1] P. Thurston, *The topology of 4-dimensional G-spaces and a study of 4-manifolds of nonpositive curvature*, PhD-thesis, University of Tennessee, Knoxville (1993).

[Thu2] P. Thurston, *$CAT(0)$ 4-manifolds possesing a single tame point are Euclidean*, Preprint, Cornell University (1993).

[ThW] W. Thurston, *The geometry and topology of three-manifolds*, Lecture Notes, Princeton University, 1978.

[Vo] L. Volkovyskii, *Plane geodesically complete subsets of spaces with nonpositive curvature*, Algebra i Analys 6 (1994), 90–100 (Russian); English translation: St. Petersburg Math. J. 6.

[Wa] A. Wald, *Begründung einer koordinatenlosen Differentialgeometrie der Flächen*, Ergebnisse eines math. Kolloquiums, Wien 7 (1935), 24–46.

[Wo1] J. Wolf, *Homogeneity and bounded isometries in manifolds of negative curvature*, Illinois J. Math. 8 (1964), 14–18.

[Wo2] J. Wolf, *Growth of finitely generated solvable groups and curvature of Riemannian manifolds*, J. Differential Geometry 2 (1968), 421–446.

[Wot1] T. Wolter, *Homogene Mannigfaltigkeiten nichtpositiver Krümmung*, Dissertation, Universität Zürich (1989).

[Wot2] T. Wolter, *Geometry of homogeneous Hadamard manifolds*, International J. Math. 2 (1991), 223–234.

[Wot3] T. Wolter, *Homogeneous manifolds with nonpositive curvature operator*, Geometriae Dedicata 37 (1991), 361–370.

[Wot4] T. Wolter, *Einstein metrics on solvable groups*, Math. Zeitschrift 206 (1991), 457–471.

[Ya1] S.-T. Yau, *On the fundamental group of compact manifolds of nonpositive curvature*, Annals of Math. 93 (1971), 579–585.

[Ya2] S.-T. Yau, *Seminar on Differential Geometry*, Annals of Math. Studies 102, Princeton Univ. Press & Univ. of Tokyo Press, 1982.

[Zi] R. Zimmer, *Ergodic Theory and semisimple groups*, Monographs in Math. 81, Birkhäuser, Boston-Basel-Stuttgart, 1984.

Index